Social and Ecological Interactions in the Galapagos Islands

More information about this series at http://www.springer.com/series/10427

Thomas Kvan • Justyna Karakiewicz
Editors

Urban Galapagos

Transition to Sustainability in Complex Adaptive Systems

Springer

Editors
Thomas Kvan
Melbourne School of Design
Faculty of Architecture
Building and Planning
University of Melbourne
Parkville, VIC, Australia

Justyna Karakiewicz
Melbourne School of Design
Faculty of Architecture
Building and Planning
University of Melbourne
Parkville, VIC, Australia

ISSN 2195-1055 ISSN 2195-1063 (electronic)
Social and Ecological Interactions in the Galapagos Islands
ISBN 978-3-319-99533-5 ISBN 978-3-319-99534-2 (eBook)
https://doi.org/10.1007/978-3-319-99534-2

Library of Congress Control Number: 2018959412

This Springer imprint is published by the registered company Springer Nature Switzerland AG
The registered company address is: Gewerbestrasse 11, 6330 Cham, Switzerland

Series Preface

In May 2011, the University of North Carolina (UNC) at Chapel Hill, USA, and the Universidad San Francisco de Quito (USFQ), Ecuador, jointly dedicated the Galapagos Science Center, an education, research, and community outreach facility on San Cristobal Island in the Galapagos Archipelago of Ecuador. The building dedication in 2011 was the culmination of an emerging partnership between UNC and USFQ that began several years earlier through a 2006 invitation to Carlos Mena and Steve Walsh to assist the Galapagos National Park and The Nature Conservancy in a remote sensing assessment of land cover/land use change throughout the archipelago. Leveraging related work in the Ecuadorian Amazon, Carlos Mena (USFQ Professor of Life and Environmental Sciences) and Steve Walsh (UNC Lyle V. Jones Distinguished Professor of Geography), Co-Directors of the Galapagos Science Center, traveled throughout the islands using preprocessed satellite imagery and spectral and geospatial equipment to validate preliminary analyses of the Galapagos with a focus on invasive plant species. Since that project, Mena and Walsh have continued to regularly engage the Galapagos Islands, coordinating research conducted at the Galapagos Science Center by faculty, staff, and students from both campuses as well as by collaborating scientists from institutions around the globe who together seek to understand the social, terrestrial, and marine subsystems in the Galapagos Islands and their linked and integrative effects. Now with nearly 50 permitted park projects operating at the Galapagos Science Center and a diversity of scientific topics being studied using a host of theories and practices, innovative work continues in an assessment of compelling and vital ways. The state-of-the-art facilities at the Galapagos Science Center include nearly 20,000 ft^2 of space that supports four laboratories (i.e., Microbiology and Genetics, Terrestrial Ecology, Marine Ecology, and Geospatial Modeling and Analysis), operated by a permanent administrative and technical staff, to support science, conservation, and sustainability in the Galapagos Islands. In addition, students enroll in classes taught by UNC and USFQ faculty as well as conduct research to complete their undergraduate honors theses, graduate theses, and doctoral dissertations. And several scientists at the Galapagos Science Center engage the community on topics including water and pathogens, nutrition and public health, tourism and community development, marine ecology and oceanography, and invasive species.

From these beginnings and with the general intention of developing a Galapagos Book Series to document our findings, highlight special needs, and describe novel approaches to unravel the social-ecological challenges to the conservation and sustainability of the Galapagos Islands, the Springer Book Series was launched through its inaugural book, *Science and Conservation in the Galapagos Islands: Frameworks & Perspectives*, edited by Steve Walsh and Carlos Mena and published by Springer in 2013. The series has continued to expand, with books, for instance, on evolution, the Galapagos Marine Reserve, and Darwin and Darwinism, energy mix on islands, and understanding invasive species in the Galapagos Islands. Now with considerable pleasure we welcome, *Urban Galapagos: Transition to Sustainability in Complex Adaptive Systems*, edited by Tom Kvan and Justyna Karakiewicz, University of Melbourne, Australia.

The general goals of the Galapagos Book Series are to examine topics that are important in the Galapagos Islands but also vital to island ecosystems around the globe. Increasingly, viewing islands as a coupled human-natural system offers a more holistic perspective for framing the many challenges to island conservation and sustainability, but the perspective also acknowledges the important context of history, human population, migration of plants, animals and people, economic development, social and ecological disturbances, and the evolution and adaptation of species (including humans) on islands to changing circumstances. This book offers new and compelling insights and further adds to the Galapagos Book Series in important ways.

Chapel Hill, NC, USA — Stephen J. Walsh
Quito, Ecuador — Carlos F. Mena

Contents

List of Contributors

Jaime Eduardo López Andrade Universidad San Francisco de Quito USFQ, Colegio de Arquitectura y Diseño Interior, Quito, Ecuador

Michael Batty Centre for Advanced Spatial Analysis (CASA), University College London (UCL), London, UK

Luís M. A. Bettencourt Mansueto Institute for Urban Innovation, Department of Ecology and Evolution, Department of Sociology, University of Chicago, Chicago, IL, USA

Santa Fe Institute, Santa Fe, NM, USA

Mark Burry Smart Cities Research Institute, Swinburne Research, Hawthorn, VIC, Australia

Camilo Cruz Melbourne School of Design, Faculty of Architecture, Building and Planning, University of Melbourne, Parkville, VIC, Australia

Fjalar J. de Haan Melbourne School of Design, Faculty of Architecture, Building and Planning, The University of Melbourne, Parkville, VIC, Australia

Kim Engie Department of Geography, Center for Galapagos Studies, University of North Carolina at Chapel Hill, Chapel Hill, NC, USA

Paola A. Espin School of Biological and Environmental Sciences, Universidad San Francisco de Quito, Ecuador and Galapagos Science Center, San Cristobal Island, Galapagos, Ecuador

Diego Quiroga Ferri Universidad San Francisco de Quito USFQ, Colegio de Ciencias Sociales y Humanidades, Quito, Ecuador

Brian G. Frizzelle Carolina Population Center, University of North Carolina at Chapel Hill, Chapel Hill, NC, USA

Justyna Karakiewicz Melbourne School of Design, Faculty of Architecture, Building and Planning, University of Melbourne, Parkville, VIC, Australia

Geoff Kimm Melbourne School of Design, Faculty of Architecture, Building and Planning, University of Melbourne, Parkville, VIC, Australia

Michael Kirley School of Computing and Information Systems, University of Melbourne, Parkville, VIC, Australia

Thomas Kvan Melbourne School of Design, Faculty of Architecture, Building and Planning, University of Melbourne, Parkville, VIC, Australia

Carlos F. Mena School of Biological and Environmental Sciences, Universidad San Francisco de Quito, Ecuador and Galapagos Science Center, San Cristobal Island, Galapagos, Ecuador

Philip H. Page Center for Galapagos Studies, University of North Carolina at Chapel Hill, Chapel Hill, NC, USA

Francesco Pizzitutti School of Biological and Environmental Sciences, Universidad San Francisco de Quito, Ecuador and Galapagos Science Center, San Cristobal Island, Galapagos, Ecuador

Duke University, Durham, NC, USA

Stephen J. Walsh Department of Geography, Center for Galapagos Studies, University of North Carolina at Chapel Hill, Chapel Hill, NC, USA

Chapter 1
Complexity and Consequence in Coupled Natural Urban Systems

Thomas Kvan and Justyna Karakiewicz

We start by stating the obvious: human activities have profound impacts on the environment. While there are more apparent and singularly evident mechanisms by which we harm the natural environment, such as through pollutants and waste, the impact of human settlements is more extensive yet less examined outcome. It has been widely noted that our human population is moving increasingly into urban or collocated settlements, concentrating our footprints into ever larger towns and cities. Such colocation brings economic and cultural benefits but concentrates environmental impacts. There is increasing work that recognizes that urban and natural ecosystems must be considered as continuities and that zoning regulations that prescribe areas for settlement do not delineate boundaries of these systems.

The Galapagos Islands are an appropriate living laboratory in which to consider both the consequences of a growing urban population and the interlinked systems. The resident population of the islands is growing as are the number of visitors who wish to see the distinctive ecology first-hand. The permanent and transient human population of the islands is growing, posing obvious and real challenges to the primary source of the opportunity, the natural environment. The chapters of this book therefore examine matters that have consequences for sustainable cities by taking a case study of a particular and distinctive example.

Only 5 out of 18 major Galapagos Islands are inhabited with permanent human settlements. Until the Galapagos National Park (GNP) was established in 1959, both settled and natural areas were managed in the same way. The GNP was assigned 97% of the total land area, and the two areas were subject to distinct management through zoning. The boundaries between National Park (natural) and inhabited

T. Kvan (✉) · J. Karakiewicz
Melbourne School of Design, Faculty of Architecture, Building and Planning,
University of Melbourne, Parkville, VIC, Australia
e-mail: tkvan@unimelb.edu.au; justynak@unimelb.edu.au

T. Kvan, J. Karakiewicz (eds.), *Urban Galapagos*, Social and Ecological Interactions in the Galapagos Islands, https://doi.org/10.1007/978-3-319-99534-2_1

areas are carefully maintained to control the detrimental impact of human settlements on natural ecosystems.

While the primary focus of research in the islands has been on the natural systems, this Springer series has been established to examine human-environment interaction both for consequential outcomes but also to inform better management of these. Other volumes have examined social and economic aspects of human activity; this volume turns attention to the forms and means of habitation. Any human presence must be supported by urban settlements as the presence is contained to limited footprints. Supporting life in these settlements requires consumables and transportation, generates waste and introduces species as well as pollutants.

Short of banning all access to the islands, it must be acknowledged that both areas are essential to the future of the islands. The National Park cannot operate without staff as managers and guides; thus both natural and urban areas are essential components. The delineation is, however, artificial. Species, flora and fauna, cross the boundaries. The border is thus not a distinct line between two systems; our understanding of the islands will benefit from a closer consideration of the two as linked and codependent systems that are both areas of protection and areas of production (Valdiva et al. 2014).

The Galapagos Islands are a living laboratory for considering the impact of human settlements, and this includes postulating the alternatives for future actions. This volume engages the framing of complex adaptive systems theory to consider the interactions of human settlements in the mutually dependent situation in the islands. In doing so, it moves beyond common practice in urban design and planning that relies on zoning and simple model descriptions. We have become increasingly aware of the limitations and the inadequacy of using "concepts tied to logical positivism such as rationalism, reductionism and comprehensive long-range planning to address the problems and challenges of the urban environment" (McAdams 2008). We now perceive that our previous tools in considering social change, an assumption that every observable effect has an attributable cause or that complicated phenomena can be understood through by breaking the complex problem into smaller coherent pieces to analyze separately or that sufficient analysis of the past events can provide the capability to predict future events, are irrelevant and misleading (Jones and Hughes 2003).

While complex systems theory has been present in many disciplines for the past 40 years, such as economics, management, biology, and social science, the majority of urban planning and design has followed a problem-solving paradigm. This approach often leads to an intensification of the problem or the creation of consequent problems through inappropriate interventions. To quote Forrester (1969), *we misunderstand symptoms for causes*. He notes that complex systems are counterintuitive since causes cannot be found in prior events but rather in "the structure and policies of the system." Therefore, even when we speak about complex adaptive systems (CAS) in the urban context, we remain trapped in simple systems thinking, identifying linear causality. It has been noted that this condition is present in the Galapagos:

> The Management Plan of the Galapagos National Park (GNP 2005) recognizes that an enormous effort has been invested in the last three decades to attack the effects and not the real causes of the problems… To be able to effectively change the current path of Galapagos we must look at the archipelago from different perspective, one that will permit the identification of real causes of the crisis. (Tapia et al. 2009)

What is absent in the discussion to date is a framework in which to align our understanding or the urban and natural systems and a framework in which to identify ways forward to a sustainable future for the community resident on the islands. Key to the transition is the displacement of fishing as the primary economic driver in the island economy by a rapidly growing tourist economy. Most tourists visit for boat-based tours in which little of the income reaches the island population, thus the growth as little effect locally.

This is changing, however, to capture locally more of the benefits of tourism. Local skills and services need to be created to ensure the growth is not only in the boat traffic but also contributes a land-based tourism in which residents can participate. The challenge is therefore identifying how this might be developed without adding to the problems created by a reliance on products and produce shipped from the mainland, as is described in chapters of this volume, and to identify a transition strategy by which a more resilient future can be realized. A component in this transition will be reconsideration of the urban forms to manage the growth in population without a commensurate expansion of the urban footprint nor the negative consequences of consumption.

In this volume, we have brought together researchers working with CAS theory in both natural and human systems. The work here is the outcome of a workshop held in October 2016 at the Galapagos Science Center in Puerto Baquerizo Moreno on San Cristobel island in the Galapagos that came about from discussions between Justyna Karakiewicz, Carlos Mena, and Stephen Walsh in which they recognized that their work, respectively, in urban and natural research drew upon a common underpinning, CAS theory. This offered a framework in which to consider the implications and opportunities of population growth on the islands.

The following chapters are organized to lead the discussion from the background to human settlement to move into the theoretical tools to understand the context. Current conditions are then examined to illuminate the challenges and to understand what has occurred both top-down through regulation and bottom-up through emergent activities. These insights are then applied to postulate how further change might occur, using the CAS concept that interventions can lead to productive change.

The volume starts by providing the background to understanding the urban challenges of the islands and progresses to set out the theory to understand these opportunities. In Chap. 2, Jaime López and Diego Quiroga describes the physical condition and situation of the islands focussing on the history of their urban settlements and the general parameters of the natural and urban ecosystems. They note that the urban areas are the loci of most human activity on the islands and also where most invasive species and contamination occur. Regulations have not been effective in containing human activity and the inefficient use of resources, particularly land surface, makes the problem worse than it might be. Noting that urban settlements are necessary but negative influences, the authors call for a deeper understanding of these areas and their role in the environment.

In Chap. 3, Michael Batty, Luís M. A. Bettencourt and Michael Kirley then introduce CAS theory and its application in urban settlements. Rather than conceiving of such settlements as static, the authors use CAS to consider the dynamic nature of urbanism. From this, they develop from a general modeling framework to characterize coupled urban natural systems, extending the characterization presented in coupled human and natural system (CHANS) models. The coupled urban natural systems framework provides insight to guide urban development that delivers more resilient outcomes which, in turn, can inform sustainability planning in the Galapagos.

The path to a sustainable future requires change and that change will take place in a global context. Fjalar de Haan, Diego Quiroga, Stephen Walsh, and Luis Bettencourt turn our attention in Chap. 4 to the transformative change that must occur to move a society from the current unsustainable trajectory and explore how urban centers can be drivers for positive change. The authors consider the change of scale at which these drivers act to enable socio-economic transition. The changes that have occurred in the island can be understood as part of larger systems of change and also that the transitions that occur in a place may be consequences of needs elsewhere. As they note, "the place in transition is not necessarily where the transition 'comes from'." Addressing the question whether growth in the urban population might be a driver for positive change, they frame this in complexity theory to consider self-organizing and co-evolutionary change to postulate how such change might occur and argue that the next transition should be essentially *galapagueña*, that is, a distinctive and locally defined transition.

The following chapters apply these approaches to current conditions and provide a foundation for exploring change. Chapter 5 by Stephen J. Walsh, Kim Engie, Philip H. Page, and Brian G. Frizzelle considers the changes that have occurred in the Galapagos as ecotourism has grown in value, and the declaration of natural parks has limited the economic activity of the fishing population. The authors employ the framework of a coupled natural-human systems to develop an agent-based model of fishing and tourism and explore five scenarios of change on the islands. They note that social, administrative, and organizational systems have failed to keep pace in the change, revealing points of failure that have exacerbated problems in the balance of human and natural systems. The chapter concludes with reflections on implications more generally for agent-based models and complexity theory in human-environmental interactions.

The next chapter looks more closely at the regulatory systems. Diego Quiroga provides a more detailed history of human settlements on the Galapagos Islands in Chap. 6, in particular at the development of regulatory controls. He documents the changes in population and activity over time, with the consequent risks to the environment that underpins the economy. In this history, he reveals how top-down and emergent processes have worked together to transform the islands and their environment, documenting in this the complex adaptive system that has operated there and the opportunities for intervention within the system.

The major economic driver in the island systems is tourism and any consideration of future opportunities for transitions to sustainability must address this economy. Paola A. Espin, Carlos F. Mena, and Francesco Pizzitutti present a systems dynamics model for tourism in Chap. 7 to articulate in particular the dimensions of sustainability for the islands. The model illuminates the mechanisms by which the tourist industry influences the islands and their economy and provides particular insights to the questions of scale and feedback in the system. The social and economic implications of tourism strategies are explained as are the environmental dangers of an ever-growing tourism economy to sustain unconstrained population growth, a path that can lead to collapse.

The last two chapters bring design approaches to bear in understanding how the preceding material can be interpreted to inform urban change. In Chap. 8, Justyna Karakiewicz takes these insights into a design-led thought experiment to consider how urban development can be the agent of positive change. She uses Holling's cycle of adaptive change and panarchy to develop a model of the natural, urban, and knowledge cycles and then applies these to inform specific strategies for urban interventions. The chapter considers specifically the opportunities in the challenges inherent in the handling of waste, the management of invasive species and the provision of potable water and illustrates interventions using material from design studios held in San Cristobel between 2014 and 2017. These challenges are addressed by changes to urban form and economic activities and postulate a *galapagueña* resilient urban future.

A design approach can be used to enable a broader engagement in decision-making to effect these transitions in urban form. Mark Burry, Camilo Cruz, and Geoff Kimm generalize the design approach in Chap. 9, outlining how computational methods in design might be used more generally to support end-user engagement and communication into design action. The model postulates a decision-making approach in the generation of architectural form by considering the specificity of human settlement needs in the Galapagos Islands and the conditions of the complex adaptive systems as described in earlier chapters. A proof of concept system is presented using a reactive scripting environment.

Together the chapters in this volume provide a theoretical insight to a sustainable island future, consider examples of change that are occurring and postulate how interventions in urban settlements might be realized through design.

References

Forrester JW (1969) Urban dynamics. MIT Press, Cambridge

Jones W, Hughes SH (2003) Complexity, conflict resolution, and how the mind works. Confl Resolut Q 20(4):485–494

McAdams MA (2008) Complexity theory and urban planning. Urban Affairs and Policy, IX. Chicago, Urbana

Tapia W, Ospina P, Quiroga D, Reck G, Gonzales JA, Montes C, Cruz E, Milstead B, Wolff M, Watkins G, Carrion C, Piu M, Oviedo M (2009) Toward a shared vision of Galapagos: the

archipelago as a socioecological system. In: Galapagos report 2007–2008. GNPS, GCREG, CDF and GC, Puerto Ayora, Galapagos, Ecuador, pp 11–15
Valdivia G, Wolford W, Lu F (2014) Border crossings: new geographies of protection and production in the Galapagos Islands. Ann Assoc Am Geogr 104(3):686–701

Part I
The Galapagos as a Living Laboratory

Chapter 2
The Galapagos Urban Context

Jaime Eduardo López Andrade and Diego Quiroga Ferri

Introduction

The Galapagos are an archipelago, and the inhabited areas are widely dispersed although there are three primary settlements. For this reason, human impact such as traveling from one island to the other has a much larger effect on the islands beyond the inhabited zones. Due to the nature of the archipelago's configuration, inhabitants build more infrastructure than if it were just a single concentrated settlement.

This chapter will further explain the relationship between the inhabitable areas in the Galapagos and its demography through an elaborate description. The main sources of data for this discussion are the plan for protected areas of Galapagos 2014 and Galapagos Census 2010 (CGREG 2012; Instituto Nacional de estadísticas y Censos 2010), unless otherwise noted. Additionally, the chapter uses the 2010 cadastral GIS maps provided by the municipality of Santa Cruz and relevant data obtained in the field by the University of Melbourne and University San Francisco de Quito.

The chapter provides a description of the Galapagos Islands as context for later chapters. First, a general description of the Galapagos built environment presents the reader with an overall view of all inhabited islands by stablishing the main facts and relationships. Second, it introduces the areas of more intense human activity in the archipelago, showing the development between the main relationships in the

J. E. López Andrade (✉)
Universidad San Francisco de Quito USFQ, Colegio de Arquitectura y Diseño Interior, Quito, Ecuador
e-mail: jlopez@usfq.edu.ec

D. Quiroga Ferri
Universidad San Francisco de Quito USFQ, Colegio de Ciencias Sociales y Humanidades, Quito, Ecuador
e-mail: dquiroga@usfq.edu.ec

T. Kvan, J. Karakiewicz (eds.), *Urban Galapagos*, Social and Ecological Interactions in the Galapagos Islands, https://doi.org/10.1007/978-3-319-99534-2_2

most inhabited islands. Third, it further describes the built environment of the islands of Santa Cruz and San Cristobal. The conclusion does not offer a closure of the study but opens up the door for further analysis in relation with the subject. All tables and graphics are elaborated crossing references from the information described above (Fig. 2.1).

General Description of the Galapagos Built Environment

The Galapagos Islands have a population of 25,124 people distributed over three urban and five rural settlements.[1] Out of the 234 land masses (Consejo de Gobierno del Regimen Especial Galapagos 2015), only 5 islands contain human settlements and their infrastructure. Santa Cruz, Baltra, San Cristobal, Isabela, and Floreana islands are located around the geometrical center of the archipelago. In the article "The geographic opening of Galapagos," Christopher Grenier referred to the islands of Baltra and Santa Cruz as the core of the archipelago because most of the human interaction is happening between the inhabited island of Santa Cruz and the airport located in Baltra (Grenier 2009). In his article, Grenier did not consider San Cristobal to be part of the core of human habitation; according to Grenier, the fact that San Cristobal is the capital of the archipelago is an inheritance from the past. However, with the increased number of commercial flights arriving to San Cristobal, due to the renovation of the airport in 2014 and the new institutional buildings and infrastructure built in the last decade, San Cristobal has gained significant activity and needs to be considered as part of this core (Fig. 2.2).

Santa Cruz has an extension of 98,522 ha with an inhabitable area of 11,610 ha. It is the most populated island, with 15,393 inhabitants that represent 61% of the population of the entire province (Table 2.1). Located in the center of the archipelago, it has three human settlements, the rural areas of Santa Rosa and Bellavista and the urban area of Puerto Ayora. These are connected by a main highway that links the airport on Baltra Island located to the north of Santa Cruz, with the town of Puerto Ayora located at the south. There are two ports, one in the north by Baltra called Itabaca Canal and one in Puerto Ayora called Academy Bay, which receive boats from inhabited and touristic islands. Santa Cruz has the biggest volume of inland tourism of all the islands.

Baltra Island covers 25,370 ha. From 1942 to 1949, it was occupied by the "Beta Base" of the American army to protect the Panama Canal during the Second World War (La Torre 1999). Located at the north of Santa Cruz, it contains the main airport

[1] "Urban areas are the zones with the highest concentration of built structures where most human activities take place, collection and distribution centers of people and products, but also areas with the biggest amount of changes in the environment due to constructions, introduced species and contamination. Rural areas are zones devoted to farmland activities such as agriculture and cattle, generally located in the elevated areas of the inhabited islands, but also clearly associated with the environmental effects derived from farmland." Dirección del Parque Nacional Galapagos (2014).

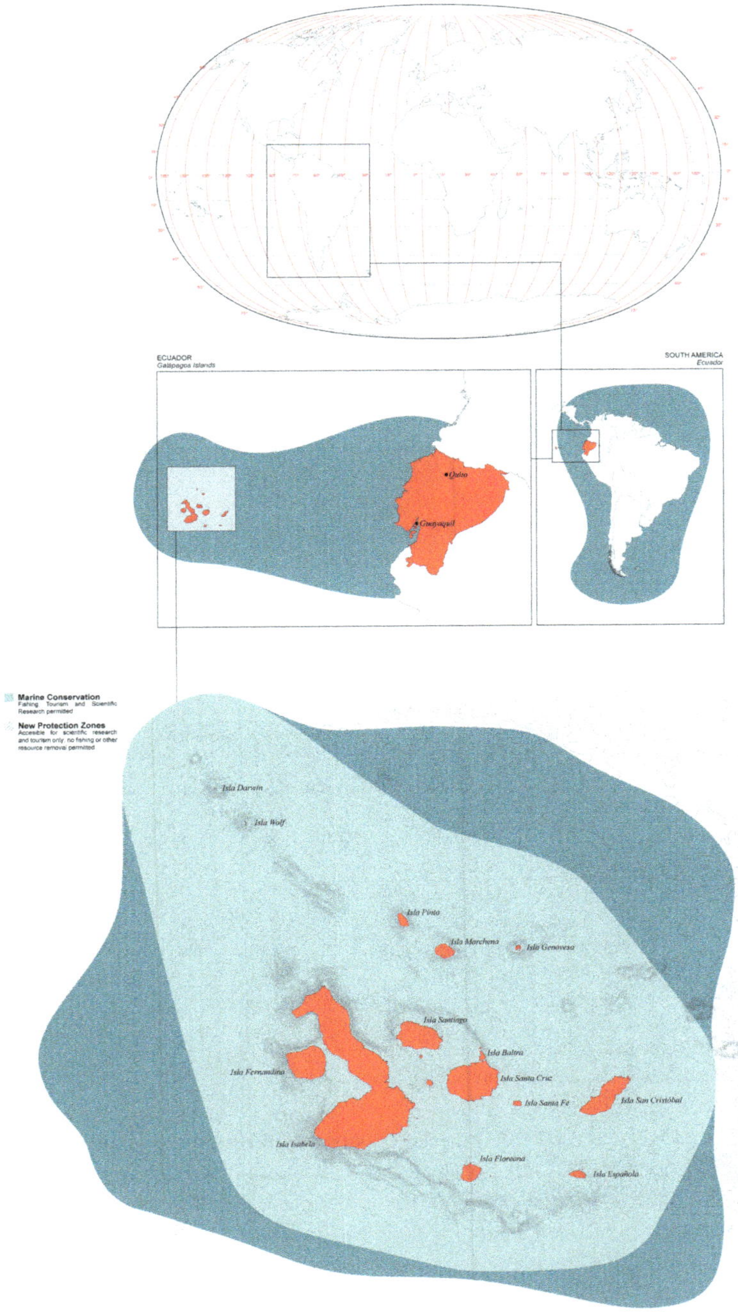

Fig. 2.1 The location of the Galapagos Islands

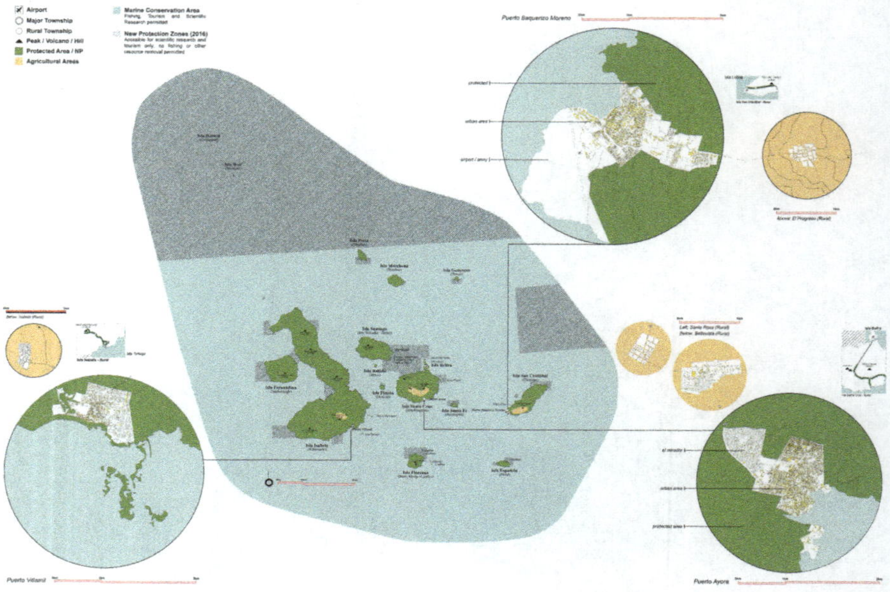

Fig. 2.2 The Galapagos Islands

Table 2.1 Inhabitable areas and population in the five islands of human interaction

Island	Total area (ha)[a]	Urban area (ha)	Rural area (ha)	Inhabitable area (ha)	% of inhabitable area	Population	% of the population
Santa Cruz	98,522	190	11,420	11,610	44%	15,393	61%
Baltra	2,537	0	0	0	0	0	0
San Cristobal	55,697	730	8,340	9,070	34%	7,330	29%
Isabela	470,344	110	5,210	5,320	20%	2,256	9%
Floreana	17,229	0	310	310	1.2%	145	0.6%
		1,030	25,280	26,310		25,124	

Source: CGREG (2012)

[a]Source: Dirección del Parque Nacional Galapagos (2014)

of the archipelago with 201,000 passengers coming through per year (Ministerio de Turismo del Ecuador 2014). It is connected with Santa Cruz through Itabaca Canal.

San Cristobal has an area of 55,697 ha with an inhabitable area of 9,070 ha. It is the second most populated island with 7,330 inhabitants, 29% of the population (Table 2.1). Located at the east of the archipelago, it is the island closest to the continent. There are two human settlements, one identified as rural and one as urban. There is an airport with a traffic of 70,000 passengers per year (Ministerio de Turismo del Ecuador 2014) and one port that connects with Puerto Baquerizo Moreno.

Isabela is the biggest island of the archipelago; it has an extent of 470,344 ha with an inhabitable area of 5,320 ha. Located at the west, it has a small population of

2,256 inhabitants divided between the rural area of Tomas de Berlanga and the urban area of Puerto Villamil. It has one airport for interisland flights of small aircrafts, as well as one port, both located close to the town of Puerto Villamil. Originally colonized in the late nineteenth century, from 1945 to 1959, it was home of a prison community (La Torre 1999). Today, its main economic driver is tourism.

Isla Floreana has an extension of 17,229 ha, with an inhabitable area of 310 ha all of which are declared as rural. It is located at the south of the archipelago and contains the smallest population of the province with 145 inhabitants. This island was occupied from 1833, by Villamil who was sent by the Ecuadorian government with small groups of some 200–300 people, who were seen and described by Charles Darwin during his 1825 visit. According to the data published on the book *La Historia Humana de Galapagos* (La Torre 1999) and the census of 2001 and 2010, its population never surpassed 400 inhabitants. It contains one port and it can only be reached by boat.

The Direction of the Galapagos National Park (DPNG) has official defined borders of the protected areas. However, legal mechanisms and negotiations have been used between the municipalities and the Galapagos National Park to allow the settlements to grow pass the already defined limits. An example of this was the negotiation known as "El Mirador" in Santa Cruz Island in 2009. This project allowed the town of Puerto Ayora to considerably extend its urban surface (Guevara 2009).

Even though each inhabited island is different from the others, their territorial structure shares some recognizable elements. Farmlands are located in the highlands and urban settlements near the sea (Valdivia et al. 2014). In the highlands, some of the rural private properties have been left abandoned. Uncontrolled farmland provokes invasive species to spread. Invasive plants such as raspberry, the guava, and the quinine tree; invasive insects such as wasps, ants, and mosquitoes; invasive animals including birds like smooth-billed ani, frogs, geckos, goats, pigs, donkeys, cats, and dogs; and many microorganisms some of which are pathogenic to the people and the endemic flora and fauna constitute the most important biological threat to the islands.

Of all the inhabited islands, Santa Cruz and San Cristobal contain the most concentrated human habitation. These two islands represent 90% of the entire population of the archipelago. At the same time, they also contain 79% of the inhabitable territory of the Galapagos (Table 2.1). Almost all the interaction between the Galapagos and mainland Ecuador occurs on these islands. One third of passengers arrive through the airport of San Cristobal and two thirds through Baltra to connect to Santa Cruz (Ministerio de Turismo del Ecuador 2014). All products that travel from the mainland by boat sail via Guayaquil to San Cristobal, Santa Cruz, Isabela, and Floreana. They off-load 95% of their freight in San Cristobal and Santa Cruz (Consejo de Gobierno del Regimen Especial Galapagos 2015).

Owing to its geographic location, the island of Santa Cruz has become the preferred tourist destination for people looking to take day tours to other islands. It has the largest concentration of tour boats and taxi boats, 40 in total, including those in Puerto Ayora and Canal Itabaca (Consejo de Gobierno del Regimen Especial Galapagos 2015). Although Puerto Baquerizo Moreno, on San Cristobal, remains comparatively small, it is historically the capital of the province and has grown substantially over recent years due to governmental investments. These investments facilitated the development of infrastructures to serve the entire archipelago. Furthermore, it has shown an increase in land-based tourism operations. To fully understand the human impact on the archipelago, it is necessary to review these two islands on a comparative and complementary study.

Santa Cruz Island and the Town of Puerto Ayora

Santa Cruz was the last island to be colonized. It was settled in the 1940s by Europeans, North Americans and Ecuadorian colonizers. It was also visited by the California Academy of Science expedition in 1906. It is from the Academy, the name of the boat that carried this scientific expedition, that the bay nearby what is now Puerto Ayora gets its name. In the late 1950s, Santa Cruz was chosen as the ideal location to establish a research center, the Galapagos National Park headquarters. This was accompanied by the creation of travel itineraries and the identification of places that allowed visitors. Because of these different institutions, the town of Puerto Ayora became a dynamic and rapidly growing urban center, strategically located in the core of the archipelago. The fast increase in the number of arrivals, from only 5000 tourists a year in the early 1970s to more than 220,000 in 2017, and the fact that many of the operations started to be based in Puerto Ayora meant that Santa Cruz grew without an urban and socioeconomic plan. As the area around the town densed, new neighborhoods, including El Mirador, and developments, such as hotels, restaurants and other facilities for tourists, started to emerge in the highlands.

On the island of Santa Cruz, there are three distinct settlements, one classified as urban, Puerto Ayora, and two as rural, Santa Rosa and Bellavista. Puerto Ayora hosts 11,974 inhabitants; it is the biggest settlement on Santa Cruz and the most populated area in the archipelago. The second largest settlement on Santa Cruz is Bellavista. With a population of 2,425 people, Bellavista is the third most populated settlement in the Galapagos. Furthermore, Bellavista has shown a rapid growth on its population between 2010 and 2015 (Table 2.2), and it is evolving to become a satellite town of Puerto Ayora, and several areas of urbanization are emerging between the two concentrated urban masses (Fig. 2.3).

According to the government council, the town of Puerto Ayora has an area of 263 ha; 79 of them were added in 2009 (Consejo de Gobierno del Regimen Especial Galapagos 2015). This extension, known as El Mirador, remains sparsely inhabited, although there are roads, a few houses, and hotels. Based on the 2010 census, Puerto Ayora reached a density of 63 inhabitants per hectare in an area of 190 ha (CGREG 2012) that excludes El Mirador.

As stated in 2010 GIS cadastre, in Puerto Ayora, the sum of private and public plots reaches a total of 146 ha of buildable urban surface, without considering roads and pedestrian infrastructure. The town has a built area of 38.07 ha in the ground plane, which gives ground plane utilization of 26% of the buildable area. Furthermore, if we include all different floors, buildings reach a total of 59.94 ha of

Table 2.2 Galapagos population 2010, 2015

Island	Settlement	Total population 2010[a]	Resident population 2010[a]	Resident population 2015[b]
Santa Cruz	Pto. Ayora	11,974	10,999	11,822
	Bellavista	2,425	2,386	3,384
	Santa Rosa	994	513	495
	Total	**15,393**	**13,898**	**15,701**
San Cristobal	Pto. Baquerizo Moreno	6,672	6,140	6,553
	El Progreso	658	654	535
	Total	**7,330**	**6,794**	**7,088**
Isabela	Pto. Villamil	2,092	2,001	2,164
	Tomás de Berlanga	164	164	180
	Total	**2,256**	**2,165**	**2,344**
Floreana	Floreana	**145**	**136**	**111**

Source: CGREG (2012)
[a]Total population refers to all the people living in the Galapagos in a determined moment, being either national or foreigners, regular or irregular. Resident population refers to the people that live in the Galapagos for more than 6 months
[b]Source: Instituto Ecuatoriano de Estadísticas y Censos (2015), INEC (n.d.)

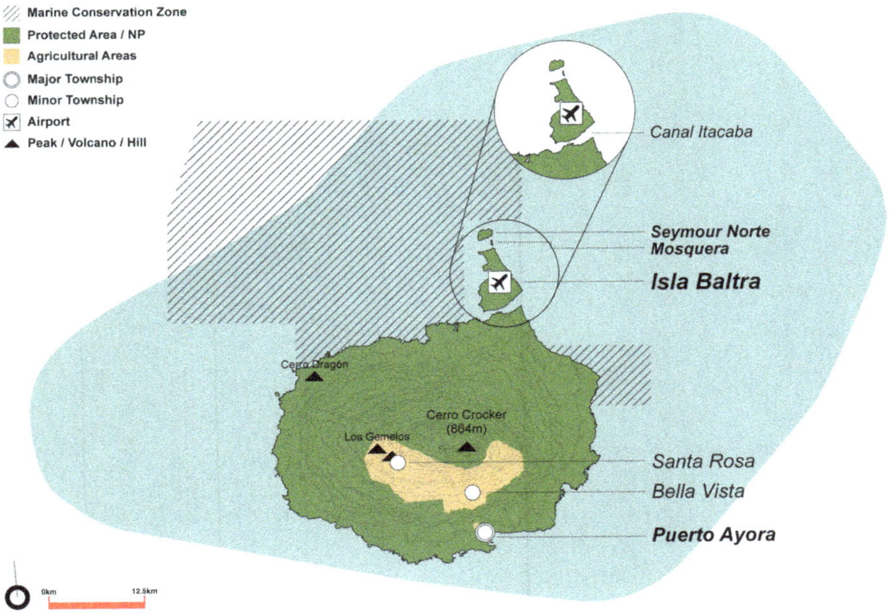

Fig. 2.3 The island of Santa Cruz

building construction, giving an occupation ratio of 41% of the buildable surface (Table 2.3). This type of occupation ratio shows that the intensity of use of space in Puerto Ayora is still very low, even though people have a perception of high density;

Table 2.3 Use of urban space, comparative data between Puerto Ayora and Puerto Baquerizo Moreno

	Puerto Ayora	Puerto Baquerizo Moreno
Concentrated urban area (SqM)	1,802,360	1,323,843
Buildable area (SqM)	1,469,285	869,411
Ground floor built area (SqM)	380,680	248,747
Ground floor occupation ratio (%)	26%	29%
Total built area	599,422	378,176
Total occupation ratio (%)	41%	43%

Source: GIS cadastral map 2010, municipality of Santa Cruz, database USFQ

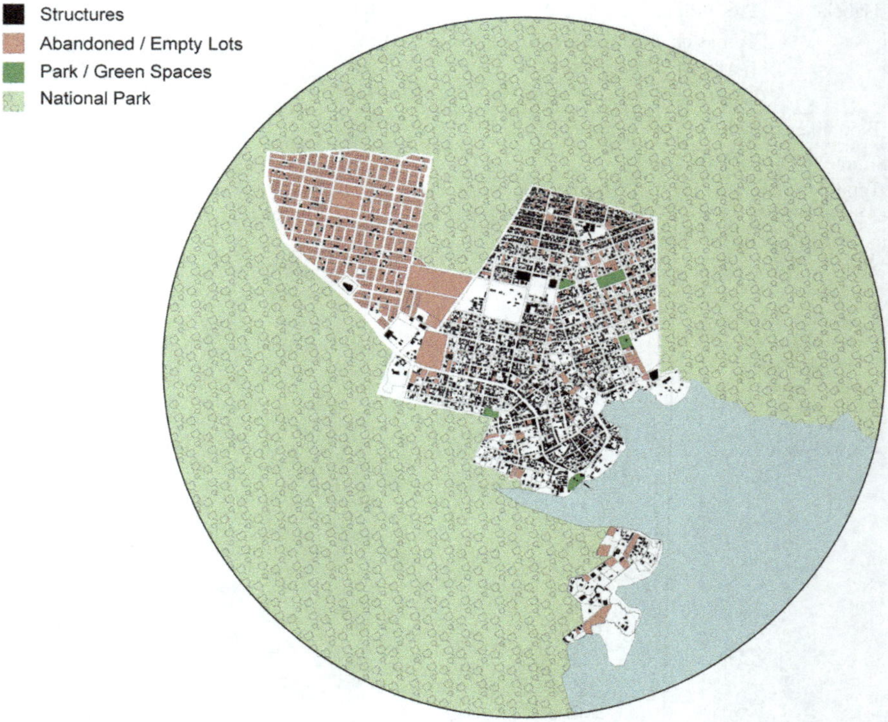

Fig. 2.4 Puerto Ayora

also, there is a high provision of built space per inhabitant, equivalent to 50 SqM per capita (Fig. 2.4).

Puerto Ayora's urban structure is roughly defined by two streets, one the center of economic activity and the other a community focus. The former lies along the seafront, and the latter connects the city with the highlands and the airport. The seafront is also the largest concentration of tourism services. The community area, located in the inner city, has the biggest concentration of community buildings and housing (Table 2.4).

Table 2.4 Urban elements of Puerto Ayora and Puerto Baquerizo Moreno

The town of Puerto Ayora Puerto Ayora is located on the southwest of the island of Santa Cruz; it has a total of 263 ha (Consejo de Gobierno del Regimen Especial Galapagos 2015) on the urban area and a population of 11,974 inhabitants. Its main connections are with Bellavista in the highlands of Santa Cruz and the airport in Baltra Island	**The town of Puerto Baquerizo Moreno** Puerto Baquerizo Moreno located at the west side of San Cristobal Island; it has an area of 730 ha and 6672 inhabitants. Its main connection is with El Progreso in the highlands of San Cristobal and with the airport located at 10-min walking distance from the town
Topography Puerto Ayora does not have abrupt changes on its topography. However, the highest part of the city is 20 m higher than the lowest area. Planned areas as El Mirador did not respect natural terrain, changing the elevation by filling the topographical accidents. As a result, streets remain elevated in comparison with the plot	**Topography** The city of Puerto Baquerizo has a few changes on its topography. The north area is less elevated than the west part of the city. It has two topographical accidents; one hosts a built lookout for the landscape of a historical peer, and the other hosts the new facilities for government council
Streets Puerto Ayora has two main roads: the Charles Darwin Avenue and the Baltra Avenue. Over these two main roads, there is a concentration of commerce and services. Tourist services are found over the Charles Darwin Avenue at the seafront. Puerto Ayora's seafront is partial, because most of the land with front to the sea is either private or institutionally owned Street patterns reveal that there have been several additions to the original urban fabric, the last one being El Mirador on 2009	**Streets** Puerto Baquerizo is structured over four main roads, the Waterfront Street, 12 de Febrero Avenue, Alsacio Northia Avenue, and Juan Jose Flores Street. The Waterfront is a piece of public infrastructure that extends over the seafront. It connects the city with the piers and creates a public walk of about 800 m long. It contains the largest concentration of hotels and tourist activities Alsacio Northia Avenue is the main street in the city; it connects the entire city in N–SW direction; it communicates the airport with the city Juan Jose Flores Street runs parallel to Alsacio Northia dividing the city in half and connects several institutional buildings. 12 de Febrero Avenue is the urban extension of the highway connecting the rural area of El Progreso with the Waterfront
Public Spaces Puerto Ayora has two main public spaces, one park inside the city and one located on the seafront in connection with the pier. In both cases, these areas become part of the urban structure. Furthermore, the beach of Tortuga Bay, located at 3.5 km from the city, acts as a public space. The same thing happens with "Las Grietas" which is a body of water inside a ravine easily reached by foot. During weekends, it is a tradition for the people of Puerto Ayora to go to Bellavista Park	**Public Spaces** The main public space of Puerto Baquerizo is the Waterfront. Along this walk, there are places to sit and rest; it is well used by the community and tourists. Puerto Baquerizo also has a city park with a piece of infrastructure to make concerts and other types of presentations. However, institutional buildings are built over the park. Mann Beach is also an important public space. Sea lions are always lying on this beach, which makes it into an important tourist attraction
Other facilities Public areas on the surroundings of the Charles Darwin Research Station and La Playita are also easily reachable from the town, although these are mostly used by tourists. La Playita is a small beach located in the path to the Charles Darwin Research Station	**Other facilities** Puerto Baquerizo has an interpretation center and historical sightseeing, such as Darwin Bay close to town, which are mostly used by guided groups of tourists

(continued)

Table 2.4 (continued)

Public buildings The most recognizable public buildings are the hospital, the sports center, the community center, the energy interpretation center, and the stadium. The Charles Darwin Research Station and the Directive of Galapagos National Park are inserted inside protected areas; in both cases its infrastructure is built on the National Park territory The hospital is located over the Charles Darwin and Baltra avenues, in front of the pier. The community and sports center are inside the city, next to one of the parks. The energy center and interpretation center can be found over the road to Tortuga Bay beach	**Public buildings** Puerto Baquerizo Moreno contains many institutional and public buildings. The hospital is equipped to receive patients from all inhabited islands making it the biggest health facility of the archipelago. Other administrative buildings that serve the entire archipelago are civil registration, emergency central, and the main facilities for the Galapagos government council GCREG. Other important buildings are the airport and the stadium. Finally, Puerto Baquerizo Moreno hosts the Galapagos extension of the University San Francisco de Quito and the Galapagos Science Center, a research station run by the same university in agreement with the University of North Carolina, Chapel Hill
Neighborhoods The different areas of the town are easily identifiable by their forms, streets, plots, and building aggregation patterns that define the different neighborhoods. In some cases, it is clearer than others, as it happens on "La Cascada" neighborhood at the northeast of the city. It has a very specific size of plot; streets' width differs from the rest of the town. Besides, one area comes apart from the concentrated urban mass, and it is reachable only by boat	**Neighborhoods** Puerto Baquerizo Moreno structures around a green core generated by a public park, some private facilities, and an educational institution. The differences on the street patterns in the area that surrounds this core divide the town in different identifiable zones. The zone that is more easily recognizable is the one immediately adjacent to the seafront. Its street pavement changes, the streets become more pedestrian friendly, and it is widely used by tourists and locals
Function Puerto Ayora can be appreciated as the economic hub of the archipelago; it has a bigger presence of tourism and tourist agencies than the other islands. The major concentration of tourist activities is near the seafront; however some hotels and restaurants can be found very much inside the urban area. It also has an important concentration of commerce and services toward Baltra avenue. Apart from this, housing concentration areas are easily recognizable around neighborhood such as "Matazarno," "La Cascada," and "El Belen"	**Function** Puerto Baquerizo Moreno is the political and administrative capital of the archipelago. However, it also has a very important tourism activity; the difference in the tourism facilities with Puerto Ayora is that at a primary observation in Puerto Baquerizo Moreno, formal hotels seem to be more present than informal ones. The major concentration of tourism activity is at the seafront. Institutional activities can be found over one of the main axes; the rest of the city is a mixture of uses with predominance of dwelling
Housing Single-family homes, even though they are predominant, are not built as a priority. In many cases, houses are built with spare rooms or in-law units for rent. In some areas, people tend to build their houses slowly, growing the construction as they have more income. As a result, a lot of constructions have unfinished columns at the top of the buildings. Many houses on Puerto Ayora seem to have been built as an informal construction. Multiunit housing is very rare; buildings with more than two floors are usually hotels	**Housing** As in the rest of the islands, housing is mostly left at its own luck. There are several empty lots on the city; few private houses show architectural response to the environment. It is difficult to find examples of multiunit housing. As it happens in other islands, people prefer to build their homes with extra rooms or floors to host tourists

More than half of the population lives in single-family homes, while the remaining live in rented units inside single houses or other types of dwellings (CGREG 2012). Most buildings are of one and two levels, and only 9% of all buildings are taller than two floors.

Even though Puerto Ayora has a considerably larger population density compared to other islands, the facts above show how inefficiently urban space is being used and explain that the extension of "El Mirador" was unnecessary. Also, these facts present an opportunity to not only avoid further growth but also to rethink the urban area of Puerto Ayora into a more efficient distribution of space.

San Cristobal Island and the Town of Puerto Baquerizo Moreno

San Cristobal was first colonized by Manuel J. Cobos in the later part of the nineteenth century. He established a large sugar and coffee plantation and a cattle farm. Cobos brought modern machinery from the USA and Europe to process the sugar. Cobos coined his own money and had the only ship that brought goods from the mainland and took the products of the Hacienda of El Progreso. After Cobos was assassinated by his own workers who considered him a brutal boss, San Cristobal became an island that depended mostly on the fishing and agricultural production. Since Puerto Baquerizo is the capital of the Galapagos, it is the base for local government institutions and the home of many public servants both residents of the Galapagos and also people coming from the mainland. Recently tourism has become a very important sector. In San Cristobal and Isabela, more land-based operations attract *scuba* divers, sport fishermen, and surfers. These are new types of tourism activities. The town is now also becoming an important center for academic activities with the establishment of the extension of the Universidad San Francisco de Quito and the Galapagos Science Center. Several NGOs have created and attracted many international volunteers that work on restoring ecosystems.

The island of San Cristobal has two human settlements, one urban, Puerto Baquerizo Moreno, which is the capital of the province, and one rural, El Progreso. The town of Puerto Baquerizo Moreno has a population of 6,672 inhabitants; it is the second biggest and second most populated settlement in the Galapagos. The other settlement of San Cristobal is El Progreso with a population of 658 people. The rural area by El Progreso is farmland, much of which is abandoned; El Progreso had suffered a decrease in its population between 2010 and 2015 (Table 2.2, Fig. 2.5).

The town of Puerto Baquerizo Moreno has a defined urban area of 730 ha. Even though it has the largest urban area in the archipelago, only one fifth of it is occupied; the rest belongs to the navy and remains empty (Consejo de Gobierno del Regimen Especial Galapagos 2015). The concentrated urban area comprises 153 ha (Consejo de Gobierno del Regimen Especial Galapagos 2015), suggesting a density of 44 inhabitants per hectare. To calculate the use of space, this study did not con-

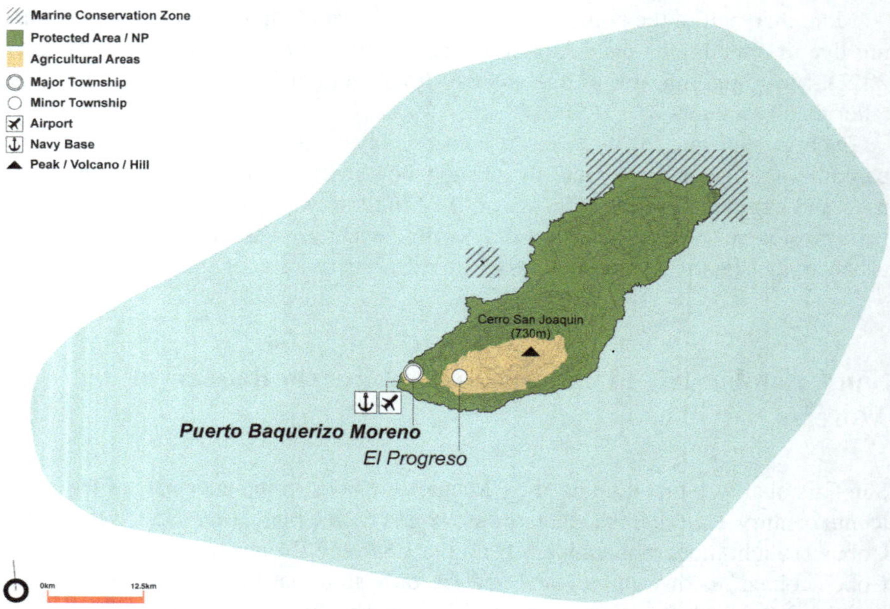

Fig. 2.5 The island of San Cristobal

sider the empty area that belongs to the navy or the plots along the highway that goes to the highlands, which are not urbanized yet.

The town has 86.9 ha of buildable surface in all private and public plots and a built area on ground floor of 24.87 ha, which gives a ground floor occupation ratio of 29%. Furthermore, including all the built floors, it has 37.8 ha of building construction, giving an occupation ratio of 43% of the buildable surface (Table 2.3). An urban area with such a low ratio of occupation becomes inefficient in the use of different types of infrastructure. However, it registers a high ratio of built space over the number of inhabitants; the amount of construction is equivalent to 56 SqM per capita (Fig. 2.6).

The town structure can be roughly defined into four areas, separated by four streets (Table 2.4). There are an area of intense tourist activity, an area used by multiple types of institutions, a primarily residential area, and an area that has mostly commercial activities. Out of the four streets, three cross the city lengthwise, each connecting important elements of the city such as the airport, public buildings, or the seafront. The remaining one runs perpendicular connecting the city to the rural area (Table 2.4). One- and two-story buildings are the most common types of construction; only 10% of buildings are taller than two floors. According to the census, almost half of the population of Galapagos lives in single-family homes (CGREG 2012). Institutional buildings are formally planned. However, there are entire neighborhoods that are poorly built, without taking into consideration the environment or the landscape in which they are located.

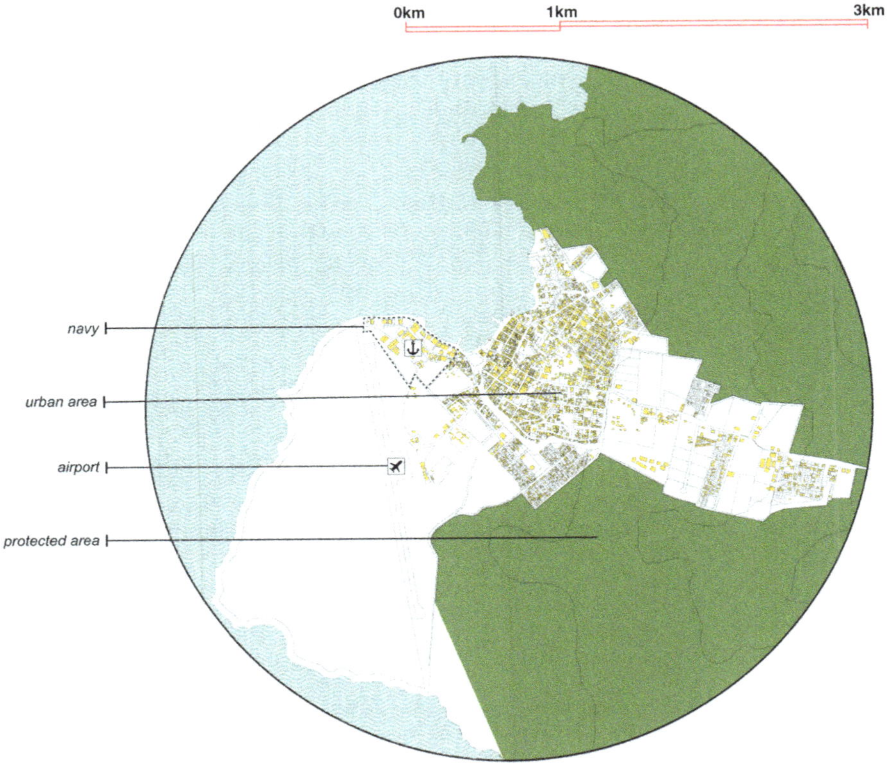

Fig. 2.6 Puerto Baquerizo Moreno

Conclusion

The Galapagos has spurred several studies in the areas of biology, evolution, geology, and many more sciences. During the last few years, human inhabitance and ecosystems have been a key focus of institutional research. It is mandatory to start seeing the Galapagos holistically. Institutions in the Galapagos such as the Galapagos National Park and Charles Darwin Foundation study biological issues such as evolution and later conservation from a biological perspective. It is only later in the late 1980s and 1990s that there is any real research about the economic, social, and cultural situation of the islands. Nevertheless, human inhabitation has played a big role in the built environment; it is an area that has hardly been researched.

In order to first approach this issue, a descriptive analysis of Galapagos human habitat was needed. This study will help contextualize the necessity for further studies of Galapagos built environment and increase the awareness of its impact in the archipelago. The Galapagos population is fast growing and will continue to do so. It is mandatory for the sake of the islands to understand the urbanization processes.

There are two islands with the most development and inhabitation. In both cases, the inhabited areas have already passed 10% of the area of the island. The built environment has been an important part of the islands for many years. The two main towns show clear signs of pressure for growth, despite their low density and inefficient use of space.

Urbanization can become a destructive force in the islands if its process is not understood and guided. With better understanding, we may be able to mitigate its impact and accommodate inevitable growth in the population in a less destructive urban form. It is essential though to start with a good understanding of the current condition.

References

CGREG (2012) Principales características demográficas de Galápagos – Resultados censo 2010. Talleres Gráficos del CGREG, Puerto Baquerizo Moreno, Galápagos, Ecuador

Consejo de Gobierno del Regimen Especial Galápagos (2015) Plan Galápagos 2015–2020, vol I. Registro Oficial, Puerto Baquerizo Moreno, Galápagos

Dirección del Parque Nacional Galápagos (2014) Plan de manejo de áreas protegidas para el buen vivir. DPNG Puerto Ayora, Galápagos

Grenier C (2009) The geographic opening of Galapagos. The Galapagos Report, pp 121–129

Guevara C (2009) La construcción de una sociedad sustentable. Universidad de Cuenca, Cuenca, Ecuador

INEC (n.d.) Instituto nacional de estadisticas y censos. Retrieved from censo de poblacion y vivienda galapagos 2015. http://www.ecuadorencifras.gob.ec/documentos/web-inec/Poblacion_y_Demografia/CPV_Galapagos_2015/

Instituto Ecuatoriano de Estadísticas y Censos (2015) Principales resultados censo de población y vivienda Galápagos 2015. INEC, Quito

Instituto Nacional de Estadísticas y Censos (2010) Resultados del Censo 2010, fasciculo provincial Galápagos. Instituto Nacional de Estadísticas y Censos, Quito

La Torre O (1999) El Hombre en las Islas Encantadas. FUNDACYT, Quito

Ministerio de Turismo del Ecuador (2014) Boletin de estadísticas turísticas. MINTUR, Quito

Valdivia G, Wolford W, Lu F (2014) Border crossings: new geographies of protection and production in the Galapagos Islands. Ann Assoc Am Geogr 104(3):686–701

Chapter 3
Understanding Coupled Urban-Natural Dynamics as the Key to Sustainability: The Example of the Galapagos

Michael Batty, Luís M. A. Bettencourt, and Michael Kirley

Introduction

The Galapagos archipelago is arguably the world's best-known example of a series of naturally evolving ecosystems where human intervention has been minimal and reasonably controlled since sustained human contact began over 500 years ago. With Darwin's landing in the 1830s and the popularization of the fragile but charismatic ecosystems described in his chronicles (Darwin 1839), interest in visiting the islands intensified. Nevertheless, it was only much later, as international travel became easier, that large-scale ecotourism took off.

Over the last few decades, increasing economic revenues from tourism have become important to Ecuador's economy, with the consequence that the number of visitors to the islands is growing exponentially and is now beyond 250,000 each year. Despite migration controls, a corresponding growth in the local population and the development of the built environment on land is affecting life on the islands, including the introduction of new species and new materials, thus increasing the pressure on ecosystems from pollutants on the land, air, and water.

M. Batty (✉)
Centre for Advanced Spatial Analysis (CASA), University College London (UCL), London, UK
e-mail: m.batty@ucl.ac.uk

L. M. A. Bettencourt
Mansueto Institute for Urban Innovation & Department of Ecology and Evolution, University of Chicago, Chicago, IL, USA

Santa Fe Institute, Santa Fe, NM, USA
e-mail: bettencourt@uchicago.edu

M. Kirley
School of Computing and Information Systems, University of Melbourne, Parkville, Australia
e-mail: mkirley@unimelb.edu.au

T. Kvan, J. Karakiewicz (eds.), *Urban Galapagos*, Social and Ecological Interactions in the Galapagos Islands, https://doi.org/10.1007/978-3-319-99534-2_3

To better understand these pressures and to formulate a plan toward mitigating or eliminating them over time, we need to begin characterizing the evolution of these coupled natural and human systems. Although there is an increasing attention to our understanding of the couplings between human and natural systems, our knowledge of these dynamics is still quite rudimentary (Liu et al. 2007). It consists largely of rather simple conceptualizations of linkages between both kinds of system that are strongly asymmetric. Human socioeconomic dynamics tend to be seen as driving forces, characterized by a different speed and level of complexity, and understood differently from most passive natural ecosystem dynamics. The dynamic adaptive relations between human and ecosystem dynamics tend to be portrayed in rather simplistic terms at best, typically abstracting away processes of human development and natural evolution, and are barely known, at worst.

In this chapter, we will sketch what we know about such systems as a preliminary to starting our speculations about how the Galapagos Islands might best adapt and deal with the pressures on the rich ecosystems that constitute their nature. We will begin by sketching what we know about complex adaptive systems in general before launching into the linkages, couplings, and balances that define how such systems coevolve. This, we suggest, is a prelude to defining a series of models that we can be taken forward in thinking about how we might explore the workings of these systems.

A Brief History of Systems Thinking

There are two very different classical approaches to systems thinking with respect to man-made and natural systems. Man-made systems are manufactured. In the past, the practice of these systems has often treated them as being intentionally created and centrally organized – from the top down – such as in modernist urban planning. The analogy with machines, as well as an emphasis on efficiency, has been widely developed with respect to this engineered systems approach.

For natural systems, the approach to their structure and dynamics is based primarily on ideas of evolution by natural selection, that is, on selection acting on differential growth, a process that acts from the bottom up. Both types of system and their domains might look as though they can be explained using either approach and therein lies much confusion. Our starting point will be from man-made rather than natural systems largely because through the last century, the top-down systems approach has slowly given way to the bottom-up approach in which human systems resemble systems that grow and evolve rather than being planned from the top down. In short, the idea that a man-made system is more like a machine is slowly giving way to one that treats it as more like an ecosystem, and in that sense, there is a convergence between natural and man-made systems and an opportunity to approach their coevolution that we will exploit here.

The notion that a relatively well-defined physical artefact or natural object constitutes a bounded system whose components function mechanistically goes back many centuries. However, in the 1920s, a distinct movement began to consider somewhat less well-defined systems in science and to a lesser extent in social science using the machine analogy. Some sciences, particularly biology, had not then developed much beyond the stage of rudimentary classification, while others such as psychology were shrouded in so much mystique that any attempt at their objectification and formalization was regarded as essential clarification. In this milieu, von Bertalanffy (1950) coined the term general system theory to describe a system, which was an actively functioning mechanism distinct from its wider environment where the environment acted benignly with respect to the import and export of energy. Rashevsky (1938) provided a mathematical framework for many such systems, while Lewin (1951) speculated on a more holistic frame for taking such systems theory into the human realm, on the way rephrasing the term first used by Koffka in Gestalt psychology: "the whole is other than the sum of its parts."

These movements and their advocates bifurcated into two main traditions, which today have become the modern theories of engineering and of complex systems. On the one hand, engineered systems increasingly took the form of mechanisms that could be controlled to achieve some desired purpose, such as a steam engine or a thermostat. In such systems, there is a controller separate from the system and its environment, which adaptively steers the system to some predetermined state reflected in the system's goals. After World War II, developments in electrical engineering and in basic computing then led quickly to what Wiener (1948) called cybernetics – the "art of steersmanship" – that provided many operational tools for the applications of these ideas to physical and social systems alike. Today, control theory is the main general theory of engineering, operating sophisticated machines, from nuclear power plants to self-driving cars and computer systems. The point about controlled systems is that they tend to perform specified functions and thus be adaptable – in a stationary sense – only within a narrow envelope of conditions.

The second systems science tradition deals what has become known as complex systems. Although often complex systems entail controllable, efficiency-driven systems in engineering, they cast a much wider net toward understanding the open-ended mechanisms of learning and adaptation that give rise to evolution by natural selection, ecosystem dynamics, human organizations, and cities, among others.

This distinction has evolved gradually as the nature of different systems has become better understood. Over the last few decades, fields such as cognition (Park and Friston 2013), sociology (Wirth 1938), political science, economics (Krugman 1995; Arthur 1999), management science (Nelson and Winter 2004; Acs and Audretsch 1987; Auerswald 2008), and urban planning (Jacobs 1961; Alexander 1987) were all articulated, at least in part, using this "complex systems approach." These social sciences were permeated with mechanistic thinking where a strong positivist ethic was established through the application of formal models testable against empirical data and relevant to various processes of decision-making involving prediction. In many cases, such systems were conceived as being organized hierarchically from the top down, composed of interacting subsystems (modules) and their components, in a simple recursive manner (Simon 1962). They were usually perceived as structures at a cross section in time with an internal dynamic that was at best stationary and often static. The concept of negative feedback was central to restoring these systems to equilibrium, but there was little sense that such systems might evolve to a new equilibrium or might be ever out of equilibrium. Stability in structure and a corresponding sense of absolute efficiency were the key organizational principles to their dynamics (Simon 1962; Christaller 1933; Lösch 1978; Batty 2006).

The realization that such an approach is limited to static structures led in turn slowly but surely to the consideration of different structures for complex systems, still based on networks but foregoing the requirements of hierarchy and simple efficiency. This in turn would lead to the fundamental concepts of network effects, increasing returns to scale and path dependence (Arthur 1994, 2011; Jacobs 1961; Auerswald et al. 2000; Bettencourt 2014), which are all relevant for open-ended complex systems, including cities. Interestingly this approach to open-ended complex systems first started taking place in the early 1970s both among urbanists (Jacobs 1961, 1970, Alexander 2002; Alexander et al. 1977; Lynch 1984) and a variety of scientists interested in cognition and computation.

The emergence of relational analysis particularly with respect to the network structure of the brain led to learning models of neural networks, which tied attempts to generate artificial intelligence. Related ideas as to how evolutionary dynamics could be implemented computationally lead to the fertile field of so-called genetic algorithms. The work of the Michigan Group encapsulated these ideas with Burks (1971) and then Holland (1975) developing models of computable intelligence through basic developments in the dynamics of growth in cellular automata and the genetics of selection through computable algorithms. As dissatisfaction with the systems approach increased due to its focus on static mechanisms in ordered sequences from the top down, the notion that most systems of any import were not designed in this way but evolved from the bottom up has become central to our theories. Evolution gradually penetrated systems thinking, and many perspectives were then molded together to form the edifice we now call complexity theory. This in turn led to groups around the world – often convening in places such as the Santa Fe Institute – developing systematic approaches to evolutionary dynamics, as it applies to disciplines such as economics, cognitive sciences, ecosystems, and human social systems. These ideas have continued to change and have increasingly attained a level of abstraction that allows us to separate "genetic evolution" in biology from other processes, especially in cognition and human social systems, which have different detailed dynamics.

Ideas of complexity in urbanism have a somewhat separate history that is just recently connecting with ideas of evolutionary computation. Among the first few modern thinkers, the contribution from Jane Jacobs (1961) was pivotal. Reacting to the hierarchical, industrial approach of modernist planning (Le Corbusier 1986, 1987) and its tangible failure to create good cities for people, Jacobs sought to reconstruct a conceptual approach to cities based on the human experience and social networks embedded in the detailed nature of urban built spaces. As the epilogue of her most famous book, Jacobs proposes that we think of cities in terms of what she calls "organized complexity." The concept had been previously raised by

Weaver (1948), a mentor of Jacobs, who popularized Claude Shannon's information theory (Shannon and Weaver 1975), and is clearly patent among other leading urbanists, inspired by vernacular architecture and urban design, especially Patrick Geddes (1915) and Lewis Mumford (1989) (who both preceded Jacobs) and Christopher Alexander and Kevin Lynch. Much of the recent progress in developing mathematical urban theory describing cities as open-ended, coupled social and infrastructural networks can be seen as a formalization of these ideas, in close and continuing contact with new data from around the world.

Complex adaptive systems (CAS) provide the exemplars for this theory, and as a first step in examining how such theory might be applicable to the coupling of the rapid urbanization taking place in the Galapagos and the interaction between human populations and their natural ecosystems, we will now outline the essential elements of a complex adaptive system and point to how it can be considered as a template for organizing our research.

Defining Complex Adaptive Systems

Complex adaptive systems (CAS) like all systems are composed of parts whose interactions determine how the system functions in the short term and evolves over the longer term. A good definition is that their parts are sufficiently autonomous that they cannot be described by any average (of their parts), a point made by Weaver (1948) in his seminal paper where he defined three kinds of physical systems – simple deterministic systems akin to classical physics; statistical systems such as thermodynamic processes where probabilities, randomness, and averages dominated; and biological systems characterized by innovation and creativity through adaptation. As Holland (1992, 1995) emphasizes, CAS maintain their "coherence under change" and, in this sense, are recognizable under even quite radical transformations, especially related to growth and scale. It is in this sense that they evolve, and, like all systems composed of relatively autonomous parts, they function from the bottom up, each individual part or element acting purposively with respect to its own dynamic.

As CAS evolve in the sense of biological organisms, rarely being designed from the top down in the manner of many machines, some origin is always required, implicitly or explicitly. Assuming the origin, which might be an elemental part, is small and the system evolves through growth, then the way in which the system evolves – its path – is dependent on the growth process. In real systems, which grow in the "wild," this path dependence cannot be assumed to be deterministic for there is always noise and indeed constraints imposed from the environment on how the system can grow. Thus, "history matters," and no single evolving system is exactly the same as any other. This is particularly resonant with our ideas about cities: they have a spatial structure which usually relates to their origin of growth – often the market place where exchange takes place and people come together – and then rings of different land uses form around this center which are usually determined by the competition required to be as accessible as possible to the center. But there the similarity of structure ends with no one city exactly like another, and some depart radically from this spatial norm. In the Galapagos, the emergence of settlements follows this pattern on the key islands, which now illustrate an embryonic pattern of urbanization related to the predominant influences from tourism reflecting transport accessibilities and key ecological sites.

There are two additional features of a complex system that we need to emphasize. The first is that such systems are anticipatory, and by this we mean that those who comprise the urban system – its various populations – have foresight in that they can act intelligently and responsively to change. In short, they "second guess" the future and very often are involved in its design. This is a feature of complex systems that is extremely difficult to model, for such action takes place at the level of the individual but makes itself apparent at the level of the collective. In short, the impacts of foresight, as well as of adaptation to extrinsic events, generate a kind of emergence that is the product of many individual forces. The second feature is one that pertains to innovation either from inside or outside the system. Complex systems appear to get more complex as they evolve. Certainly, this is clear from an examination of ecosystems and animal populations, but it also appears to be the case for our man-made organizations and procedures that we develop so we can function socially and economically in cities. New technologies never completely replace those they seek to displace but are layered on top of previous ones and leave a strong residual signature. In this sense, cities get more complex, and they get richer and more diverse, and, in this sense, they become qualitatively different. This is something that the short history of settlement on the Galapagos clearly demonstrates from its form as a few scattered villages, housing some hundreds of population prior to the 1960s to the 25,000 or more who live in the embryonic small towns that comprise the settlements today.

The problem of human sustainable development was originally formulated in rather general terms, involving how entire societies were linked to regional and global ecosystems. Over the last decade, it has become increasingly clear, however, that towns and cities play a particularly important role in the relationship between humans and the Earth's natural environments. This is because urbanization is now a global phenomenon; nearly 60 percent of the population now live in cities, and this is predicted to increase to virtually the entire global population by the end of this century (Batty 2011). Cities also create their particular socioeconomic dynamics, through the amplification and acceleration of human interactions, which in turn are associated – locally and globally – with greater energy and resource use. Thus, human activity in cities drives demands for natural resources and controls most of their global and regional flows. In this way, processes of human development and economic growth, so necessary for fulfilling human aspirations to a better life, create a number of unintended consequences for the planet's natural environments, including biodiversity loss and ecosystem degradation due to overharvesting, pollution, and climate change. In the Galapagos, we see these processes etched out in stark relief as human populations come into increasing contact even in highly protected ways with some of the most isolated and fragile ecosystems on the planet.

The point, however, is that while energy and resource use are necessary to cities and human societies, their impacts can vary enormously and are susceptible to the same processes of innovation and adaptation as cities themselves.

Here we will address these issues by calling for a special attention to cities as complex adaptive systems. We will develop the concept of *coupled urban-natural systems* as a more specific – and we believe – more actionable outgrowth of the earlier concept of *coupled human-natural systems*, for addressing issues of sustainability in urbanizing societies. We also propose that island coupled urban-natural systems are rather special in that they create conditions where interactions and flows between social and natural systems become clearer, more easily measurable, and typically also more critical for the environment. Thus, the Galapagos can provide us with critical examples of these processes, which should lead us to understand and mitigate the consequences of urbanization in a charismatic ecosystem which as we have implied is currently experiencing explosive human settlement growth. We propose that studying island coupled urban-natural systems provides a special analytical opportunity for science and policy to build on the historical importance of these sites for natural history due to their unique ecosystems.

So much for the context. The rest of this chapter is organized in three sections. We will first provide a brief foray into the history of ideas about human settlements and cities as complex adaptive systems. We will then introduce the idea of coupled urban-natural systems and contrast them with earlier frameworks for coadaptation between human and natural environments. The following section emphasizes what is special about island coupled urban-natural systems, with an emphasis on the Galapagos. The last section discusses how data and modeling of these systems is necessary; we also will point to other specific models as being part of a general framework for sustainable development in island systems and beyond.

Cities and Ecosystems as Complex Adaptive Systems

Cities first emerged some 5,000 years ago in Asia Minor as small clusters of intense urban activity drawing their resources from the surrounding agricultural landscape. The search for a conceptual framework to understand them in all their dimensions is as old as science itself (Mumford 1989; Ortman et al. 2015). From the beginning, in classical times, the practical challenges of cities loomed large in terms of governance, health, and the environmental impacts of spatially concentrated human activity (Mumford 1989; Hall 1998; Ober 2016). Simultaneously, a fascination for the creativity that cities seemed to manifest also highlighted their central importance in the advent of new solutions to problems of the human condition and their relationship to nature (Mumford 1989; Hall 1998).

Though the history of urban planning is fraught with many attempts at idealized, instant, and top-down solutions, there is a refrain running through the work of the most important thinkers that cities are *complex adaptive systems*. This means more specifically that they are evolving networks of socioeconomic interactions between heterogeneous agents, embedded in space and time, and requiring energy, resources, and information to function. Their dynamics is nonlinear and path dependent, with many analogies but also some key differences to those of natural ecosystems, which we will elaborate below. Thus, social, economic, spatial, temporal, and infrastructural aspects of cities are interconnected and need to be understood together, in terms of adaptive dynamical systems.

In contemporary times, these ideas were articulated, mostly qualitatively, by a number of authors in the mid-20th century, including Lewis Mumford (1989), Jane Jacobs (1961, 1970), and Christopher Alexander (2002; Alexander et al. 1977). They emphasized the importance of historical contingencies, self-organization, and bottom-up, emergent dynamics as critical to the production of good urban environments. Other authors, from Patrick Geddes (1915) to Kevin Lynch (1984), also emphasized many of the similarities between the ways in which cities work and the properties of evolving ecosystems, as well as the dissimilarities with (then) existing engineered or designed systems. But only recently has more direct and comprehensive evidence on the many dimensions of cities and urban systems become available, and this is promoting a new quantitative synthesis of how cities operate and how paths to sustainable urban development may be created. Complex adaptive systems are the appropriate general frameworks to understand both the nature of dynamics of human settlements and ecosystems, as well as the emergent systems in which these interact.

Complex adaptive systems have a number of fascinating properties that make them particularly difficult to influence through traditional approaches to policy. Here we highlight five of these properties, which we will emphasize are shared also by ecosystems, but operate at different temporal and spatial scales resulting in mismatches and dynamical imbalances as these systems interact:

- First, cities, like ecosystems, are networked dynamical systems. In this way, everything in cities is connected to everything else, in the sense that social and economic networks over sufficiently long times are likely to be complete graphs, embedded in all places spanned by the built environment and infrastructure of a city. This makes the study of each agent or place in isolation problematic, as they are embedded in strongly connected networks of exchange of goods, services, energy, and information (Sampson 2012; Bettencourt 2013, 2014).
- Second, the interconnection between agents in cities allows and encourages strong heterogeneity, expressed in the deep division of labor and knowledge between people (different professions, businesses, tastes), in the advent of subcultures (Fischer 1995), and the possible coexistence of many different ethnicities, races, religions, and languages. This is a special feature of complex adaptive systems – latent also in very productive ecosystems, as in a rain forest or coral reefs – where specialization and interdependence are immense and mesmerizing (Cody and Diamond 1975; Smith 1776). The high diversity of agents is also adaptive in the sense of providing fuel for new solutions and innovations as circumstances change, for example, in terms of new economic sectors, technologies, or cultural institutions. For these reasons, cities (and ecosystems) are hard

to model, as "representative agents" (common in economic theory, Glaeser 2008) fail to capture important complementarities and synergies created by diversity and interdependence and may provide the wrong framework to assess global properties of a complex adaptive system in terms of resilience and response to the loss of particular functions.

- Third, the strong interdependence of components of a complex adaptive system typically leads to another important feature of their dynamics in terms of circular causality. For example, is a city rich because it has good infrastructure, or does it have good infrastructure because it is rich? Is someone healthier and better educated because they are more affluent or more affluent because they are healthier and better educated? This circularity of cause and effect leads to the fact that the dynamics of cities and other complex adaptive systems are best understood in terms of feedback loops, which can be positive or negative, depending on the chain of events. This feature is also shared by natural systems evolving under selection, but the means of change and time scales can be radically different.
- Fourth, as a result of their connectivity, heterogeneity, and self-reinforcing dynamics, cities beget two additional properties: emergence and open-ended growth. Emergence means that there are properties of cities as a whole that are not obvious in their parts – people or places. These are known to complex system scientists as *scaling effects* (Bettencourt 2013; Bettencourt et al. 2007a; Batty 2008) and to economists as *agglomeration effects* (Duranton and Puga 2004). They express the typical magnification of average socioeconomic outputs (and costs) with city (socioeconomic network) size and the corresponding densification of human activity over built spaces. As a result of these effects, larger cities are typically economically more productive per capita and more innovative and may – compared to smaller towns in the same system – also achieve certain economies of scale in infrastructure and energy use. This makes urbanization a pretty sure path, and a necessary condition, for economic growth and human development as we see play out globally today (Acs 2004; Arrow 1971; Barro and Sala-i-Martin 2003; Jones and Romer 2009). We can see this very clearly in the way populations are being attracted to the Galapagos despite the very strong limits on internal migration in Ecuador.
- Fifth, the last property of open-ended growth is the result of the processes of adaptation and innovation in cities (Bettencourt et al. 2007a, b; Bettencourt 2014; Acs 2004). It is itself an emergent dynamic of cities and ecosystems. However, contemporary cities and urban systems achieve growth at a typical pace of a few percent a year, a rate much too high for local ecosystems to sustain. Part of the reason for this – besides the social networked characteristics of human innovation – is that urban systems operate on much larger spatial scales This means that a modern city can harness resources not only locally, in its built area, but globally, from around the world. This concentrates energy, resources, and information to a much larger extent than in an ecosystem, which is mostly local (e.g., driven by incident sunlight or sea currents). As a consequence, the influence of a human population in a local ecosystem corresponds effectively to a large force per unit time, typically generating large disruptions that are,

even in principle, impossible for an ecosystem to match in real time. The consequence of this mismatch is that – at least over the short term – urbanization drives ecosystems much more strongly than the other way around. Managing the consequences of this drive, by eliminating or reducing its adverse consequences on ecosystems, is therefore critical and requires a finer understanding of coadaptation between human and natural systems than we currently possess.

As a result of these properties, complex adaptive systems are very difficult to plan or even influence without incurring a number of unintended consequences. Their dynamics depends to a large extent on self-organization, where agents react to each other based on local information as in an economic market or in predator-prey dynamics in an ecosystem. In such systems, the matching of physical forces and time scales is critical for self-organized stability. A mismatch between action and reaction, even in simpler engineered systems managed via control theory protocols, typically creates oscillations and instabilities that can lead to large disruptions and irreversible crashes.

For these reasons, it should now be clear that some degree of dynamic planning and top-down governance of "commons" is necessary to create good urban environments and sustainable development. The problem of better understanding the natural dynamics of cities and how to harness these mechanisms in positive directions remains one of the most important challenges for a science of cities and its applications. This challenge is further compounded when we consider the interactions of cities to natural ecosystems and seek to create sustainable outcomes in terms of forms of coadaptation of coupled urban human and natural systems. To address these issues, we propose a framework for coupled urban-natural systems, as a comprehensive approach to the measurement of the interactions and coadaptive dynamics between urbanizing human societies and natural ecosystems, with the goal of achieving paths for sustainable development across scales of analysis.

The Concept of Coupled Urban-Natural Systems

Over the last decade, the concept of *coupled human-natural systems* (CHANs) has emerged to deal with the explicit dynamics of ecosystems caused by human actions (Liu et al. 2007). This framework emphasizes the feedbacks between human societies and ecosystems, as, for example, through climate change and ecosystems disturbance, especially in terms of fisheries and agriculture, water management, and pollution. In this section, we briefly review this framework and extend it to specific issues connected to human activity in cities and urban systems.

Cities are emerging worldwide as the loci of consumption and control of most flows of energy and resources commanded by humans. For example, cities are now responsible for about 70% of worldwide carbon emissions, while at present, they host little more than 50% of the world's population. At the same time, urban areas control or capture more than 75% of the world's GDP, which indicates that cities provide the economic hubs that drive the world economy and their own nation states.[1] This concentration of demand and consumption for energy and resources plays out at many different scales and is also a feature of the Galapagos where the four or five population centers are increasingly taking on an urban character and dominate the demand for energy, construction materials, and food within the archipelago (see Chap. 8 in this volume).

Here we make the point that to better understand and affect change in coupled human-natural systems, it is necessary to understand the central role of towns and cities as drivers of their dynamics. It is our thesis that cities and urbanization, in terms of their complex and scale-dependent behavior, have not been explored sufficiently as a part of understanding the problem or devising potential solutions. To do this we introduce the concept of coupled *urban*-natural systems (CUNs) in contrast to CHANs (coupled human-natural systems) (Alberti et al. 2011). CUNs emphasize the understanding and management of flows of energy, resources, and information in and out of cities (or more generally urban systems) as the key actionable set of relations that control the relationship between human activity and ecosystems. In this way, CUNs depend on a fine, multidimensional understanding of so-called urban metabolism as first noted by Wolman (1965) which can be explored through "input-output" models and the flow systems developed in industrial ecology. These also relate to a classical tradition in geography and regional economics to understand flows between places in urban systems and associated questions of urban hierarchy and central place theory (Batty 2006).

To appreciate how coupled urban-natural systems compare and contrast with more general coupled human-natural systems, we start by listing and briefly discussing some of the properties of the latter and then emphasizing how towns and cities make their joint nature as complex adaptive systems more localized and controllable. In the well-known review we have already referred to, Liu et al. (2007) identify and discuss five properties of CHANs, which largely mirror our discussion of complex adaptive systems in the previous section. These are (1) feedback and reciprocal causal relations, (2) nonlinearity and thresholds, (3) historical path dependence and time lags, (4) resilience, and (5) heterogeneity. As we discussed above, these five properties and more are also features of cities as complex adaptive systems. They are also typically properties of any complex system evolving via mechanisms that include randomness and adaptation, including ecosystems evolving via natural (and artificial) selection.

What is most interesting when recognizing that these are features of both cities and ecosystems is to consider the whole system, resulting from their mutual coupling. In this light, it is important to recognize that in most cases urbanization endows human dynamics with the ability for much *faster* dynamics and greater energy intensity (power) than is typical of natural selection. In this sense, human dynamics connected to urbanization are the fastest, driving influence on ecosystems,

[1] We generated this percentage from applying the percent urbanized population of a country to the total GDP of the same country and expressing the sum of these GDPs as a proportion of world GDP.

driving them into disturbed regimes that may indeed be unstable or prone to collapse as their ability to respond lags. In fact, this difference in speed of adaptation (and often growth) can greatly complicate the coupling of the human and natural in that the impacts of human interventions which affect natural systems may not show themselves in the natural system for many years. Climate change is often taken as a case in point. The feedback in the opposite direction, from ecosystems to human societies, affects aspects of long-term adaptation, such as shifts in food, growth rates, and adaptations to climate. We know very little about the extent to which these environmental changes can create truly disabling conditions for human systems, which are not simply tied to local conditions, though global climate change may be an example.

A parallel issue is that once disturbed ecosystems tend to exhibit recovery rates that are slow compared to the pace of change in the human urban system, which potentially and significantly affects their resilience. This mismatch between the speed of growth, adaptation, and evolution is at the root of most unintended consequences of human activity, viz., natural environments. Moreover, both adaptation and recovery are path dependent and contingent, meaning that ecosystems even when they stop being disturbed may not recover to their precontact composition, for example, in terms of species abundances and biodiversity.

What then may be a way to create sustainable interactions between human and natural systems characterized by such different temporal and spatial scales? The way out, toward sustainable co-development, requires that influences, in speed and magnitude, of human activity on ecosystems are reduced or eliminated to the point that they do not cause degradation of the ecosystem to function over the long term. While it is hard to know what this condition may be by observing an ecosystem at any given time, it is possible to monitor species abundance, energy and resource flows, and response times over a period of time. So a close, consistent, and continuing analysis of a coupled urban-natural system is necessary that can tell us about mutual interactions and adaptation over extended periods of time and that incorporates expectations for good and bad trajectories.

In this sense, we propose the following approach that we believe is general, but it is particularly applicable to coupled urban-natural systems that can be separated in space and time from other CUNs. The first property is to view interactions not only as being balanced back and forth (in equilibrium) but typically of the human systems as controlling (in the formal sense of control theory) the natural system, as in driving it in particular ways. This is close in spirit to Odum's (1983) view of ecology as layers of systems in which energy flow and feedback are directional, a perspective that is also embedded in ideas of urban metabolism.

The second perspective grows out of this idea of energy and resource flows between urban and natural systems and asks for *innovations in the human system* that result in more synergistic exchanges, viz., the state of the natural environment and the long-term dynamics of the coupled system. This may imply efficiencies in energy and resource use so that demands from the human system are lower or generate different ways of obtaining such energy, e.g., using renewable sources which can have a dramatically smaller impact on the energy flows (including biomass) in

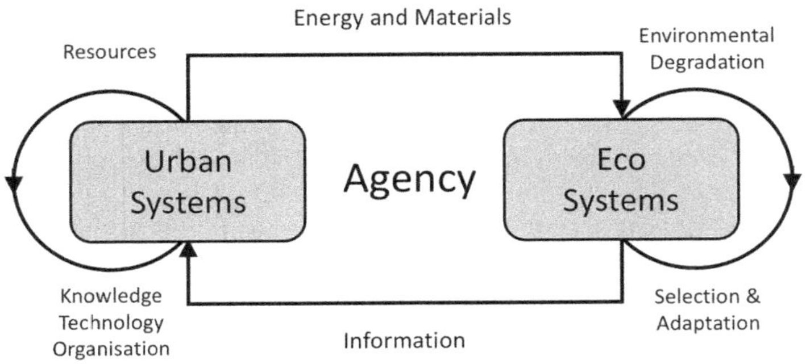

Fig. 3.1 The coupling of urban and natural systems

the ecosystems. Finally, in the spirit of control and innovation, the urban system should seek exchanges with natural systems that are either absent (being decoupled) or that are synergistic, in the sense of improving resource flows and biodiversity in the latter.

In this way, by reducing and decoupling adverse interactions between human and natural systems or more positively through the creation of synergistic exchanges, cities and human settlements may become broadly sustainable even as they continue to grow and use material resources and energy. We emphasize that this is an issue that cannot be solved through reduction in consumption alone, as cities are dissipative systems that like any other complex systems require resources to exist and generate development. Rather, the key is to seek resource flows that are sustainable, an issue that requires scientific, technical, and societal innovation, which is a major advantage of urban systems. In Fig. 3.1, we sketch the various interactions and feedback loops that provide some sense of the ways urban and natural systems both interact with one another and can generate mutually reinforcing or diverging feedbacks. We have this sort of overview in mind in the next section, which illustrates how we might begin to build mathematical models that enable us to predict conditionally what these feedbacks might be and how they will transform the Galapagos Islands.

Approaches to Modeling Urban-Natural Systems

Island economies have been modeled using many different types of approach, most of which relate economy to demography and tend to treat the physical juxtaposition of activities in different locations as addendums to the main simulations (Taylor et al. 2009). For example, a very clear and obvious approach is based on input-output modeling that articulates the local economy in terms of economic sectors which reflect to a great or lesser extension the threefold division into agriculture, manufacturing, and services – primary, secondary, and tertiary (with quaternary being a subdivision and extension of the latter). These kinds of model have been built for Galapagos by the group at University of California Davis (see Taylor et al. 2002) and focus as one might expect on tourism. We will not develop these kinds of model in any detail here for they already exist for the islands and are useful in looking at the economic consequences of tourism, but they are mainly nonspatial (Taylor et al. 2002).

We will not build any specific models in this chapter, for our purpose is to sketch approaches, but the block diagram in Fig. 3.2 gives some indication of the critical sectors and feedback loops from our preliminary understanding of how the emergent urban systems interface and are integrated with the various natural ecosystems that comprise the islands. If we assume that each one of these elements can be expressed in common metrics that can be translated into one another, then we can write these relationships in Fig. 3.2 in functional terms as follows. Let us define resident population as P, employment as E, built environment as developed volume as D, and transport interactions and volume as T, with land-based tourist population as L and boat-based as B. We can then define the natural ecosystem as N but noting that this is to be further elaborated into its sectors once such a model might be developed. Changes in these volumes of activity through time are subscripted using t, and the model could then be defined for changes in a generic quantity y associated with any one of these variables as

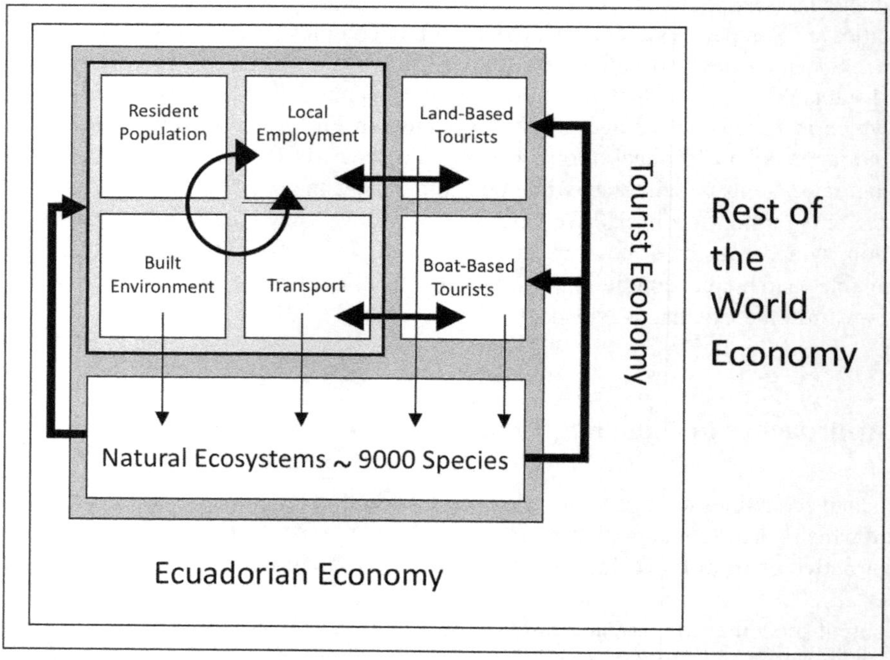

Fig. 3.2 Key components of the coupled urban-natural system in the Galapagos

$$\frac{dy_k}{dt} = \sum_j \alpha_{kj} y_j,$$

which we can expand for each of the variables we have defined as

$$\begin{aligned}
dP / dt &= \alpha_{PP}P_t + \alpha_{PE}E_t + \alpha_{PD}D_t + \alpha_{PT}T_t + \alpha_{PL}L_t + \alpha_{PB}B_t + \alpha_{PN}N_t \\
dE / dt &= \alpha_{EP}P_t + \alpha_{EE}E_t + \alpha_{ED}D_t + \alpha_{ET}T_t + \alpha_{EL}L_t + \alpha_{EB}B_t + \alpha_{EN}N_t \\
dD / dt &= \alpha_{DP}P_t + \alpha_{DE}E_t + \alpha_{DD}D_t + \alpha_{DT}T_t + \alpha_{DL}L_t + \alpha_{DB}B_t + \alpha_{DN}N_t \\
dT / dt &= \alpha_{TP}P_t + \alpha_{TE}E_t + \alpha_{TD}D_t + \alpha_{TT}T_t + \alpha_{TL}L_t + \alpha_{TB}B_t + \alpha_{TN}N_t \\
dL / dt &= \alpha_{LP}P_t + \alpha_{LE}E_t + \alpha_{LD}D_t + \alpha_{LT}T_t + \alpha_{LL}L_t + \alpha_{LB}B_t + \alpha_{LN}N_t \\
dB / dt &= \alpha_{BP}P_t + \alpha_{BE}E_t + \alpha_{BD}D_t + \alpha_{BT}T_t + \alpha_{BL}L_t + \alpha_{BB}B_t + \alpha_{BN}N_t \\
dN / dt &= \alpha_{NP}P_t + \alpha_{NE}E_t + \alpha_{ND}D_t + \alpha_{NT}T_t + \alpha_{NL}L_t + \alpha_{NB}B_t + \alpha_{NN}N_t
\end{aligned}$$

Now in this form, everything is related to everything else, and one of the key features of building a model based on this system of differential equations is the structure of the relationship matrix $\boldsymbol{\alpha}$.

$$\dot{\mathbf{p}} = \alpha\mathbf{X} = \begin{bmatrix} dP/dt \\ dE/dt \\ dD/dt \\ dT/dt \\ dL/dt \\ dB/dt \\ dN/dt \end{bmatrix} = \begin{bmatrix} \alpha_{PP} & \alpha_{PE} & \alpha_{PD} & \alpha_{PT} & \alpha_{PL} & \alpha_{PB} & \alpha_{PN} \\ \alpha_{EP} & \alpha_{EE} & \alpha_{ED} & \alpha_{ET} & \alpha_{EL} & \alpha_{EB} & \alpha_{EN} \\ \alpha_{DP} & \alpha_{DE} & \alpha_{DD} & \alpha_{DT} & \alpha_{DL} & \alpha_{DB} & \alpha_{DN} \\ \alpha_{TP} & \alpha_{TE} & \alpha_{TD} & \alpha_{TT} & \alpha_{TL} & \alpha_{TB} & \alpha_{TN} \\ \alpha_{LP} & \alpha_{LE} & \alpha_{LD} & \alpha_{LT} & \alpha_{LL} & \alpha_{LB} & \alpha_{LN} \\ \alpha_{BP} & \alpha_{BE} & \alpha_{BD} & \alpha_{BT} & \alpha_{BL} & \alpha_{BB} & \alpha_{BN} \\ \alpha_{NP} & \alpha_{NE} & \alpha_{ND} & \alpha_{NT} & \alpha_{NL} & \alpha_{NB} & \alpha_{NN} \end{bmatrix} \begin{bmatrix} P \\ E \\ D \\ T \\ L \\ B \\ N \end{bmatrix}$$

The structure of the system, as written, is linear, but this feature can be relaxed, if necessary, via the consideration of dependencies of the matrix elements on the state variables. These state variables can also be spatialized, replicating them in different locations, if desirable. The dynamics of the system depend on the structure of the matrix $\boldsymbol{\alpha}$. The number of possible relationships is clearly enormous, and a crucial step in developing these kinds of model is to hypothesize the signs of the relations and calibrate the model to gauge the extent to which our hypotheses about the dynamics of the system are appropriate.

For example, from what we know about relations in the Galapagos, we consider a first shot at these relations to be positive between population, employment, development, transport, and the land- and boat-based tourist populations and negative in diverse ways from all these growth sectors to the number of flora and fauna in the natural ecosystems. In fact, the structure of these relations is likely even with this crude set of sectors to be much more complex. Of course, the stability of these relations through time is also an issue. One of the messages of complexity theory is that as systems grow and evolve, the relations between these various sectors may change; systems may get more complex as they evolve, and this is certainly the case in the Galapagos.

Nevertheless, the point of the model above is the emphasis on defining dynamical systems where local state variables are driven by flows and interactions. Such models, to be useful, require ways to monitor and measure state variables and flows. Given models, methods of population biology and formal demography can be used to infer coupled urban-natural system trajectories, as can numerical simulations. This detailed work remains to be done but is increasingly well defined and achievable.

Discussion and Outlook

The purpose of this paper has been to provide a broad conceptual framework on cities and urbanization as complex adaptive systems and to point to the need for integrated theory and modeling of coupled urban-natural systems, a concept that we find more actionable and relevant to problems of sustainable development than previous approaches.

The central contribution from this shift in perspective is that in cities not only are they the spatiotemporal concentrations of energy and resource use in the world today, they are also the drivers for these worldwide flows and for the processes of adaptation and innovation that create them in the first place. Thus, it is by attacking the problem at its source, where there is also typically greater understanding and political will to address issues of sustainability, that we believe one has a better chance to generate the knowledge and practices that will generate sustainable development everywhere.

The discussions and work that have led to this paper were deeply motivated by the specific challenges of sustainability in the Galapagos, a charismatic and fragile ecosystem under severe pressure from settlement growth and tourism. We believe that the Galapagos and other island ecosystems are – again – ideal laboratories for the understanding and modeling of coupled urban-natural dynamics. This is because they are smaller and simpler in some ways than the world's megacities, because the flows of resources and people involved are clearer and more controllable and because its ecosystems are also more unique, pristine, and sensitive. We proposed the type of quantities that must be measured and modeled to start implementing ideas of sustainable development in the islands and look forward to contributing to a future where the most wonderful creativity of nature can coexist with that of human societies.

References

Acs ZJ (2004) Innovation and the growth of cities. In: Urban dynamics and growth: advances in urban economics, BT – Contributions to economic analysis, vol 266. Elsevier, pp 635–658. http://www.sciencedirect.com/science/article/pii/S0573855504660202

Acs ZJ, Audretsch DB (1987) Innovation, market structure, and firm size. Rev Econ Stat 69(4): 567–574. https://doi.org/10.2307/1935950
Alberti M, Asbjornsen H, Baker LA, Brozović N, Drinkwater LE, Drzyzga SC, Jantz CA, Fragoso J, Holland DS, Kohler TS, Liu J, McConnell WJ, Maschner HDG, Millington JDA, Monticino M, Podestá G, Pontius RG Jr, Redman CL, Reo NJ, Sailor D, Urquhart G (2011) Meeting report: research on coupled human and natural systems (CHANS): approach, challenges, and strategies. Bull Ecol Soc Am 92(2):218–228
Alexander C (ed) (1987) A new theory of urban design. Oxford University Press, New York
Alexander C (2002) Notes on the synthesis of form. Harvard University Press, Cambridge, MA. 17. printing
Alexander C, Ishikawa S, Silverstein M (1977) A pattern language: towns, buildings, construction. Oxford University Press, New York
Aristotle (n.d.) Politics | internet encyclopedia of philosophy. http://www.iep.utm.edu/aris-pol/. Accessed 9 Sept 2015
Arrow KJ (1971) The economic implications of learning by doing. In: Hahn FH (ed) Readings in the theory of growth. Palgrave Macmillan, London, pp 131–149 http://link.springer.com/10.1007/978-1-349-15430-2_11
Arthur WB (1994) Increasing returns and path dependence in the economy. Economics, cognition, and society. University of Michigan Press, Ann Arbor
Arthur WB (1999) Complexity and the economy. Science 284(5411):107–109. https://doi.org/10.1126/science.284.5411.107
Arthur WB (2011) The nature of technology: what it is and how it evolves. 1. Free Press pbk. ed. Free Press, New York
Auerswald PE (2008) Entrepreneurship in the theory of the firm. Small Bus Econ 30(2):111–126
Auerswald P, Kauffman S, Lobo J, Shell K (2000) The production recipes approach to modeling technological innovation: an application to learning by doing. J Econ Dyn Control 24(3):389–450
Barro RJ, Sala-i-Martin XI (2003) Economic growth, 2nd edn. The MIT Press, Cambridge, MA
Batty M (2006) Hierarchy in cities and city systems. In: Pumain D (ed) Hierarchy in natural and social sciences. Methodos series, vol 3. Springer, Dordrecht, pp 143–168. https://doi.org/10.1007/1-4020-4127-6_7
Batty M (2008) The size, scale, and shape of cities. Science 319(5864):769–771. https://doi.org/10.1126/science.1151419
Batty M (2011) Commentary: when all the world's a city. Environ Plan A 43(4):765–772
Bettencourt LMA (2013) The origins of scaling in cities. Science 340(6139):1438–1441. https://doi.org/10.1126/science.1235823
Bettencourt LMA (2014) Impact of changing technology on the evolution of complex informational networks. Proc IEEE 102(12):1878–1891. https://doi.org/10.1109/JPROC.2014.2367132
Bettencourt LMA, Lobo J, Helbing D, Kühnert C, West GB (2007a) Growth, innovation, scaling, and the pace of life in cities. Proc Natl Acad Sci 104(17):7301–7306. https://doi.org/10.1073/pnas.0610172104
Bettencourt LMA, Lobo J, Strumsky D (2007b) Invention in the city: increasing returns to patenting as a scaling function of metropolitan size. Res Policy 36(1):107–120. https://doi.org/10.1016/j.respol.2006.09.026
Burks AW (1971) Essays on cellular automata. University of Illinois Press, Urbana
Christaller W (1933) Die zentralen Orte in Süddeutschland: eine ökonomisch-geographische Untersuchung über die Gesotzmässigkeit der Verbreitung und Entwicklung der Siedlungen mit städtischen Funktionen. Gustav Fischer, Jena
Cody ML, Diamond JM (1975) Ecology and evolution of communities. Harvard University Press, Cambridge MA
Darwin C (1839) Voyages of the adventure and Beagle - capital B, volume III – Journal and remarks. 1832–1836. Henry Colburn, London

Duranton G, Puga D (2004) Chapter 48 Micro-foundations of urban agglomeration economies. In: Handbook of regional and urban economics, vol 4. Elsevier, pp 2063–2117 http://linkinghub.elsevier.com/retrieve/pii/S1574008004800051
Fischer CS (1995) The subcultural theory of urbanism: a twentieth-year assessment. Am J Sociol 101(3):543–577
Geddes P (1915) Cities in evolution: an introduction to the town planning movement and to the study of civics. Williams & Norgate, London
Glaeser EL (2008) Cities, agglomeration, and spatial equilibrium, Lindahl lectures. Oxford University Press, Oxford
Hall P (1998) Cities in civilization. Pantheon Books, New York
Holland JH (1975) Adaptation in natural and artificial systems. The MIT Press, Cambridge, MA
Holland JH (1992) Complex adaptive systems. Daedalus 121(1):17–30
Holland JH (1995) Hidden order: how adaptation builds complexity. Basic Books, New York
Jacobs J (1961) The death and life of great American cities. Vintage Books, New York
Jacobs J (1970) The economy of cities. Vintage Books, New York
Jones C, Romer P (2009) The New Kaldor facts: ideas, institutions, population, and human capital, w15094. National Bureau of Economic Research, Cambridge, MA http://www.nber.org/papers/w15094.pdf
Krugman P (1995) The self-organizing economy. Blackwell, Cambridge, MA
Le Corbusier (1986) Towards a new architecture. Dover Publications, New York
Le Corbusier (1987) The city of to-morrow and its planning. Dover, New York
Lewin K (1951) Field theory in social science. Harper and Brothers, New York
Liu J, Dietz T, Carpenter SR, Alberti M, Folke C, Moran E, Pell AN, Deadman P, Kratz T, Lubchenco J, Ostrom E, Ouyang Z, Provencher W, Redman CL, Schneider SH, Taylor WW (2007) Complexity of coupled human and natural systems. Science 317(5844):1513–1516. https://doi.org/10.1126/science.1144004
Lösch A, Woglom WH, Stolper WF (1978) The economics of location. Yale University Press, New Haven
Lynch K (1984) Good city form. MIT Press, Cambridge, MA
Mumford L (1989) The city in history: its origins, its transformations, and its prospects. A Harvest book. Harcourt, San Diego/New York/London
Nelson RR, Winter SG (2004) An evolutionary theory of economic change. The Belknap Press of Harvard University Press, Cambridge, MA. Digitally reprinted
Ober J (2016) The rise and fall of classical Greece. Princeton University Press, Princeton/Oxford
Odum HT (1983) Systems ecology: an introduction. Wiley, New York
Ortman SG, Cabaniss AHF, Sturm JO, Bettencourt LMA (2015) Settlement scaling and increasing returns in an ancient society. Sci Adv 1(1):e1400066–e1400066. https://doi.org/10.1126/sciadv.1400066
Park H-J, Friston K (2013) Structural and functional brain networks: from connections to cognition. Science 342(6158):1238411–1238411. https://doi.org/10.1126/science.1238411
Rashevsky N (1938) Mathematical biophysics: physico-mathematical foundations of biology. University of Chicago Press, Chicago
Sampson RJ (2012) Great American city: Chicago and the enduring neighborhood effect. University of Chicago Press, Chicago
Shannon CE, Weaver W (1975) The mathematical theory of communication. University of Illinois Press, Urbana
Simon H (1962) The architecture of complexity. Proc Am Philos Soc 106(6):467–482
Smith A (1776, 2003) The ealth of nations. Bantam Classics, New York. Reprint edition
Taylor J, Yunez-Naude A, Dyer GA, Stewart M, Ardila S (2002) The economics of "eco-tourism:" a Galapagos island economy-wide perspective. Available at http://reap.ucdavis.edu/research/Agricultural.pdf/Galapagos.pdf
Taylor J, Hardner J, Stewart M (2009) Ecotourism and economic growth in the Galapagos: an island economy-wide analysis. Environ Dev Econ 14(2):139–162

von Bertalanffy L (1950) An outline of general system theory. Br J Philos Sci 1:114–129
Weaver W (1948) Science and complexity. Am Sci 36:536–544
Wiener N (1948) Cybernetics: or control and communication in the animal and the machine. The MIT Press, Cambridge, MA
Wirth L (1938) Urbanism as a way of life. Am J Sociol 44(1):1–24
Wolman A (1965) The metabolism of cities. Sci Am 213(3):179–190

Chapter 4
Scales and Transformative Change: Transitions in the Galapagos

Fjalar J. de Haan, Diego Quiroga Ferri, Stephen J. Walsh, and Luís M. A. Bettencourt

Introduction

Socioeconomic transitions of regions and places are profound shifts in how local societal needs are being met – or in some cases – in how those places have been used to meet needs in other places. This latter remark already suggests that the place in transition is not necessarily where the transition "comes from." It is quite possible that a transition witnessed in a certain place is part of a process of change occurring at a larger scale that happens to *include* that particular place or that the transition is the consequence – a side effect – of occurrences pertinent to a different place,

F. J. de Haan (✉)
Melbourne School of Design, Faculty of Architecture, Building and Planning, The University of Melbourne, Parkville, VIC, Australia
e-mail: fjalar.dehaan@unimelb.edu.au

D. Quiroga Ferri
Universidad San Francisco de Quito USFQ, Colegio de Ciencias Sociales y Humanidades, Quito, Ecuador
e-mail: dquiroga@usfq.edu.ec

S. J. Walsh
Department of Geography, Center for Galapagos Studies, University of North Carolina at Chapel Hill, Chapel Hill, NC, USA
e-mail: swalsh@email.unc.edu

L. M. A. Bettencourt
Mansueto Institute for Urban Innovation & Department of Ecology and Evolution, University of Chicago, Chicago, IL, USA

Santa Fe Institute, Santa Fe, NM, USA
e-mail: bettencourt@uchicago.edu

T. Kvan, J. Karakiewicz (eds.), *Urban Galapagos*, Social and Ecological Interactions in the Galapagos Islands, https://doi.org/10.1007/978-3-319-99534-2_4

regardless of that place's scale.[1] Also, change might originate at a local scale (i.e., bottom up) but be manifested at a higher scale through the process of emergence. In such cases, the transition can be called *accidental* to that particular place. If, by contrast, the transition of a place is the consequence of local dynamics, needs, and ambitions, it is *essentially* of that place (see de Haan 2018 for the accidental-essential distinctions).

Two observations regarding scale and transitions are important to stress at this point: (1) in their geologically speaking, recent history, the Galapagos have undergone several transitions – all of them *accidental* in the above sense. The socioeconomic development of the archipelago has been determined by other places and by forces operating at different, larger scales than that of the islands themselves. These influences introduced changes that propagated through the endogenous dynamics of the islands, leading to altered social-ecological processes. And (2) the relevant scales of socioeconomic organization in general are changing. Although we are very much accustomed to the idea that the nation-state is the natural seat of power, this dominance may turn out to be a passing phase of history and, more to the point, may not apply to many issues of sustainable development. We will elaborate both these observations, but we can already prelude that the Galapagos may for the first time be able to shape a transition that affects them, making it a transition *essentially galapagueña*.

This chapter will first discuss the issue of shifting scales of relevance and review some of the insights from complexity and sustainability transition studies on this issue. We will then present a narrative of the transitions the Galapagos have gone through with the emphasis on the transition that is unfolding at the time of this writing. Then we look at how scales matter to aspects of sustainability and liveability in this transition. We conclude with some thoughts on how the Galapagos could pursue a purposive transition to sustainability making use of emerging scales of relevance.

Shifting Scales

The power to organize and change societies has been relocating. For a long while, we have assumed that the nation-state is the natural seat of power and control. The fact that the complex, technologically sophisticated and pervasive service provision systems (e.g., energy supply, agro-food, health, and transport) are typically governed at national levels is both a cause and a consequence of this as well as some economies of scale that accrue to large systems. The societal aspiration to organize certain services for everyone, the necessary engineering prowess, and the political power and mandate of the nation-states historically came together in the Second Industrial Revolution (circa the last three decades of the 19th century up until the

[1] Though one could easily argue that if a side effect is a socioeconomic transition *somewhere else*, the scale of the dynamics at its root is by implication such as to encompass that place somewhere else also

First World War). The emergence and dependency on these societal systems reinforced the importance and perceived importance of the national level.

Here, we argue that the importance of national governments for local urban sustainable development is waning and that this is just as much a historical confluence of dynamics as was its ascendance. Politically, the global scale has gained importance, but so has local scales and scale-free networks. Technologically, individuals have become empowered as they have never been, which gives rise to new alliances (e.g., sharing economies) and/or had breathed new life into old kinds of alliances (e.g., cooperatives). Culturally this is accompanied by a new kind of individualism – the keyword in many respects is now "anyone" rather than "everyone." At the same time, we acknowledge that this general trend has exceptions and counter-reactions. In our case, that of Ecuador and the Galapagos, the state is in fact actively increasing its importance and power in explicit contradiction to the ubiquitous manifestations of globalization and neoliberalization in many countries.

Against this backdrop, cities are now often considered to be the relevant scales for sustainable development. Loorbach and Shiroyama (2016), for instance, speak of cities as "transition machines." Their thesis is about cities' capacities for innovation, for being hubs of creativity. Others look to cities because nation-states seem to either lack the power or the conviction to handle sustainability issues. Indeed, cities themselves seem to take up the challenges of sustainability, resilience, carbon neutrality, etc. with a boldness and urgency that their countries tend to lack (e.g., see Hodson and Marvin 2010). States themselves may even look to cities for transitions. Hodson and Marvin (2010) note that cities may "receive" transitions as a task from national level. In these last two examples, we again see the distinction between transitions that are essentially or accidentally of a place (de Haan 2018).

Researchers certainly seem to see the potential of cities to play a leading role in sustainability transitions, judging from a recent flourishing of an apparent subfield of urban transition management, with a manual (Roorda et al. 2014) and now a book series (Loorbach et al. 2016). The idea that cities, and therefore urbanization, provide opportunities for sustainability transitions is something we will come back to in the context of the Galapagos.

Globalization can be seen as a development complementary to the increasing importance of the city scale. Sometimes, as a consequence of the diminished powers of nation-states, and sometimes causing it, supranational dynamics and institutions have become more important. Trade agreements, monetary unions, and multinational corporations can be characterized as the top-down, institutionalized part of this development, while increasing means of communication and organization (e.g., through the Internet and affordable long-distance transport) can be thought of as the bottom-up part of globalization.

These developments may also be interpreted as a development toward a situation where there is no dominant scale anymore at all. Here too, there are two distinct aspects of the problem. One is again related to increasing means of production and organization from the bottom up such as online platforms, 3D printing, etc. The open-source movement, for example, is global in the sense of not being bound by any specific geographical perimeter but is aspatial in the sense that communities agglomerate around a common topic or method without their geographic proximity being a factor.[2] 3D printing has the potential to reduce the need for – geographically binding – facilities and infrastructures, such as factories and supply chains. While this may lead to *localization* of socioeconomic activity, in combination with online networks, it means that production activities are to a large degree rid of ties to any specific geographical scale. The other aspect is reflected in the notion of "network governance." The idea here is that pursuing societal goals is no longer – if it ever were – the prerogative of governments but rather is an activity of networks, involving "a plurality of public, semipublic, and private actors" (Sørensen and Torfing 2007, p. 3).

In transition studies, the increasing importance of individuals is also being noticed. Schot et al. (2016) focus on the "user" and how she or he is no longer a passive consumer, and de Haan and Rotmans (2018) present a new theoretical framework to understand transformative change as the consequence of the actions of individual actors and alliances.

Transitions in the Galapagos

Transitions: Past

The Galapagos have gone through several transitions in their, geologically speaking, recent history. After their discovery by Europeans, the islands became a place where pirates restocked food supplies; they also served as a sort of post office for whalers. There were some early attempts at colonization since Ecuador annexed the islands in 1832. The earliest successful colonization occurred in the 1880s with Manuel J. Cobos and the Hacienda called El Progreso, creating a permanent settlement. There was also another similar settlement created in Isabela at about the same time. After these pioneering efforts, there were small waves of European settlers arriving in the 1920s and 1930s. At this point the local economy was agriculture based. In the 1950s the economy shifted toward fishing, and from the 1960s onward, tourism started developing. Fishing grew until the sea cucumber market collapsed in the mid-1990s. Tourism became the dominant industry and has grown continuously and increasingly, which in turn has driven further urbanization and population growth.

Gonzalez et al. (2008) describe the transition of the Galapagos Islands and the threats to island sustainability. Using Holling's (2001) adaptive cycle metaphor, they examine four major historical periods of the Galapagos, often punctuated by crisis events that altered future trajectories. They reference the *extractive exploitation* period (1525–1832), *colonization* (1832–1959), *wilderness conservation* (1959–1998), and *conservation-development balance* (1998–present). They also

[2] Speaking the same language, however, is often a community-binding factor in this movement. Nevertheless, in these cases also it arguably is the ideal of internationalization and accessibility that is leading.

cite three plausible future scenarios for the Galapagos – *endogenous model* "safe-keeping natural capital," *exogenous model* "consumption and stocking," and *perverse model* "sun and beach." In these scenarios, these authors highlight the current (unsustainable) tourism model in the Galapagos and some of its most critical dependencies, e.g., population growth, jobs, resource consumption, invasive species, and lifestyle of local residents. They conclude that for the Galapagos to achieve a sustainable model for the future, the islands "must modify traditional practices to produce a more adaptive resilience-based, co-management model, adopt a more comprehensive approach to territorial planning, strengthen participative approaches and institutional networks, and promote transdisciplinary research at the frontiers of social and biophysical sciences."

Transitions: Present

The human dimension in the Galapagos Islands is changing dramatically. For instance, in 1990, 40,000 tourists visited the Galapagos Islands, but the number of tourists increased to 65,000 in 2000 and 225,000 in 2016. Also, in 1990, 10,000 residents lived in the Galapagos, but today there are 30,000 residents living on four populated islands, drawn primarily to the archipelago by the burgeoning tourism and associated services industry. Nearly one-half of the population lives in the community of Puerto Ayora on Santa Cruz Island (~18,000 residents), the hub of the tourism industry, as well as in the agricultural, highland communities of Bellavista and Santa Rosa. Puerto Baquerizo Moreno on San Cristobal Island is the second largest community in the Galapagos (~9,000 residents), the seat of provincial governance and an expanding tourist destination. The agricultural community of El Progreso, the site of a historic effort by Manual Cobos in 1891 to develop a sugar factory and related agricultural enterprises, is also located on San Cristobal Island. The community of Puerto Villamil on Isabela Island (~3,000 residents) remains primarily a fishing community, but land-based tourism welcomes an increasing number of visitors drawn by unique visitation sites, the second largest caldera in the world, and a glimpse of the Galapagos Islands of 25 years ago, when the streets were sand, the environment was central, and the pace of life was slow. Finally, the community of Puerto Velasco Ibarra on Floreana Island (~500 residents) also offers touristic services and a remote and more primitive sense of Galapagos, reminiscent of the earlier years of life in the islands. The remaining islands are not inhabited by people (apart from the former military base on Baltra, now the site of one of the two airports in the Galapagos that receives flights from the mainland as well as other islands) but are visited by tourists on organized tours designed to experience the Galapagos through day trips and "island hopping" on locally owned and operated *fibras* (small speedboats that move up to 16 people and products between the islands) as well as multiday excursions lasting from 4 to 8 days on licensed boats that range from relatively small yachts to larger cruise ships, serving generally less than 100 travelers at a time. The tour boats are equipped with sanctioned tour guides, and the tourists sleep and eat onboard, occasionally visiting local communities for arranged activities and shopping. Such cruise boat-based tourism is often judged by the local population as detached and not contributing to their well-being.

Boat-based vs. land-based tourism places very different demands on the environment as well as the urban infrastructure. Land-based tourism is growing faster than boat-based tourism. Through a boom in hotel construction, the number of hotel rooms now exceeds the number of boat berths on the approximately 75 licensed tour boats. Communities are the social and economic hubs of the island, providing the majority of local jobs in tourism and to a lesser degree in government, although community infrastructure has lagged severely behind the expanding human dimension, thereby creating situations where power, water, sanitation, healthcare, transportation, and commerce are being negatively impacted. Improvements in the urban infrastructure are underway, primarily on the two main tourist centers, Santa Cruz and San Cristobal Islands. With approximately 150 sanctioned tourist sites distributed more or less equally between marine and terrestrial environments, tourist and residential visitors access sites within the Galapagos Park and Galapagos Marine Reserve to interact with iconic species, see iconic landscapes, and experience special marine environments. Increasingly, agricultural lands in the highlands of the populated islands are seeking to attract visitors to ecotourism sites that involve, for instance, the presence of migrating giant tortoises on farms and/or special geological formations such as tunnels or native and endemic vegetation such as Scalesia forests. There are also new types of activities that attract other type of tourists such as surfing and sportfishing. These activities are growing on some of the islands. Fishermen have also developed new activities such as sportfishing to try to get some of the money that is coming to the island from the flourishing tourism industry.

The promise of outstanding recreational experiences for tourists and economic opportunities for those seeking better livelihoods has led to population immigration and the "peopling" of communities in the Galapagos, primarily the coastal communities as the hub of local tourism, but also smaller agricultural communities in the humid highlands that together constitute the human use zones. These zones cover approximately 3% of the land area of the Galapagos Islands.

Residents of Galapagos are primarily drawn from mainland Ecuador and from sending or source communities, such as the two largest cities in Ecuador, Quito, and Guayaquil, as well as several secondary locations, such as Ambato and Loja. In addition, the "temporary migration" of international tourists to the Galapagos, chiefly from the United States, the United Kingdom, and Germany, is increasing as well as a significant growth in national tourism from Ecuador that has brought about increased demands on local resources that significantly change, and often stress, the social, terrestrial, and marine subsystems of the archipelago. With nearly two million visitors arriving in the Galapagos since 2000, the tourism sector directly employs nearly 70% of the residents and represents almost the entire economy (Kerr 2005; Kerr et al. 2004).

As Villacis and Carrillo (2013) report, between 2001 and 2011, the total number of tourists increased 138%, from 77,570 to 185,028, and the number of tourists who came to Galapagos in 1979 as compared to 2011 represents an increase of 1,473%, with a correlation of 0.97 between tourist and resident populations. Clearly, the human dimension in the Galapagos has been a relatively recent phenomenon with the rate of population change growing steeper with time, dominated by Santa Cruz Island, but overwhelmingly driven by the dramatic expansion in the tourism industry. Residents in the Galapagos, however, have historically been adversely affected by a poor urban infrastructure including poor educational opportunities, a deficient healthcare system, and an inadequate water and sanitation infrastructure. For instance, during the last 10 years, the Galapagos Islands have experienced an increase in the number of housing units (68%), but in terms of basic services, the Galapagos showed significant deficits, particularly in the potable water network coverage and sewer service: In 2001, only 30% of homes had network sewer service, and by 2010 this percentage decreased to 26%. These conditions, in 2010, were even worse on Santa Cruz Island, with 61% of the total population of the archipelago, with a sewage system that only covered 4% of the households (INEC 2010). With no wastewater treatment system consistently in operation in the Galapagos, wastewater is often disposed in the ocean and/or injected into the volcanic strata, thereby increasing the probability of water contamination of groundwater aquifers that are used or periodically relied upon, depending upon the island, for drinking, cooking, and bathing needs. Consequently, not only is the quality of water highly suspect, but the quantity of freshwater is also a limiting factor, especially in the case of Santa Cruz, further impacted by occasionally overpumping of groundwater systems causing saltwater and freshwater to mix, thereby introducing relatively high levels of saline water into the water distribution system as well as the mixing of wastewater and even human effluents through material injected into disposal wells with freshwater pumping sites nearby in the porous volcanic strata (Walsh et al. 2010).

Development in the communities has happened at an accelerated pace. With growing tourism and more associated jobs, migrants – legal and illegal – have arrived from the continent and/or developed locally mediated, alternate household livelihood strategies that involve part- or full-time employment in the tourism industry. Job diversification is common, certainly, by those involved in fishing and agriculture, both employment sectors that are in decline. It is common for members of households to hold multiple jobs as circumstances and opportunities dictate. As a consequence of the consolidation of jobs in the urban centers, construction of homes, as well as commercial enterprises and public services, the urban area has been intensified through both peripheral growth and urban infilling, and more recently, the verticality of urban structures is being seen, as additional floors have been added to existing structures. But, as the densification of urban structures has increased, so too has land swaps between the community and the Galapagos National Park, as formerly "firm" boundaries have transitioned to "porous" between the Park and the community, achieved through policy shifts that have also facilitated the development of new neighborhoods, formerly outside the designated human use zones.

This type of current urban development in the Galapagos does not appear to be sustainable. Presently there is no integrative strategy that links infrastructure to population and land development dynamics, where the goal is the sustainability of island ecosystems and the management of the social, terrestrial, and marine subsystems in a manner that protects the environment, supports tourism, and engages the human dimension in constructive and meaningful ways. Pizzitutti et al. (2016) offer guidance to scenario planning for tourism management, involving tourists as well as local residents, in the Galapagos, and Walsh and Mena (2016) describe the interactions of the social, terrestrial, and marine subsystems in the Galapagos Islands by viewing them as a coupled human-natural systems where feedback mechanism and nonlinear dynamics are vital for addressing "what if" scenarios of change, empowered by dynamic system models as well as agent-based models that link people and environment in fundamental ways (Malanson and Walsh 2015).

Regarding energy needs in the Galapagos, fuel arrives onboard petroleum ships that sail the 1,000 km from the continent to the islands. As the human population of the Galapagos increases, i.e., tourists and residents, additional petroleum is imported to support the consumptive demands of fuel for vehicles including, for instance, touring and fishing boats, transport vessels, and diesel fuel for the generation of electricity. Navigating shallow and irregular harbors, petroleum ships generated numerous spills, incurred during transport and off-loading. Other ships, carrying goods and products for local consumption, have run aground further exacerbating negative consequences for local flora and fauna.

Perspectives from Complex Systems

Borrowing from complexity theory and agent-based modeling (e.g., An 2012; Benenson 1999; Brown et al. 2008; Entwisle et al. 2008; Torrens 2012; Walsh et al. 2013), self-organization, emergence, and adoption and learning may be the most obvious factors associated with development in the Galapagos Islands and the close coupling of natural-human systems in this world renowned, protected area that is being impacted by the direct and indirect consequences of the expanding human dimension, with explicit implications for the environment.

One issue is the bottom-up, or informal, structure of the human occupation of the islands and its emergent properties. For years, Ecuadorians arrived in the Galapagos as tourists and stayed for jobs (often becoming illegals in their own country), now associated with tourism, but previously and less so, in agriculture and fisheries. The pull of higher-paying jobs in tourism in the Galapagos, concentrated in the communities, and the push for immigration to the Galapagos by a weak and chaotic national economy created a situation in which individuals were driven to immigration to the Galapagos resulting in the densifying of urban settings, often shaped and mediated by social networks that created important sending areas on the Ecuadorian mainland and a burgeoning population in the Galapagos linked to needed skills, experiences, and motivations. People on the Ecuadorian mainland self-organized to migrate to the Galapagos with a focus on Puerto Ayora on Santa Cruz Island, given their personal interpretation of individual and household needs, employment perceptions, wants and desires, social networks of friends and relatives having first-hand knowledge about the Galapagos or already living in the Galapagos as early migrants, and the necessity of economic gain.

Another example of complexity theory playing out in the Galapagos Islands is the adaptive behavior of fisherman into cooperatives to address the social and ecological uncertainties, primarily caused by exogenous dynamics associated with regional/global variability in commodity prices, local regulations linked to harvest quotas and seasons for lobster, sea cucumber, and white fish and alternate household livelihood strategies associated with job diversification and the implications of ENSO (El Nino Southern Oscillation) events in disrupting marine conditions, specifically depressing biological productivity through a curtailment of normal upwelling and the infusion of warm water into the Galapagos Marine Reserve. Emergence and adaptive behaviors may be observed by both the fishermen and the Galapagos National Park, who regulate fishing activities in the Galapagos Marine Reserve. Through a complex regulatory environment, the Park has created anger and uncertainty among fishing communities, generating "spontaneous" demonstrations and other forms of social unrest. Fishermen have, on occasion, fished illegally for economic gain and to register disgust with Park administration of the Marine Reserve. Among the adaptive behaviors has been the creation of fishing cooperatives, adaptive co-management of the Marine Reserve, and the diversification of individuals and households into part-time fishing or into tourism. In most cases these adaptations have been bottom up, and the interactions are direct and indirect, as agents act through intermediate agents. Individual differences among fisheries and Park managers and administrators matter, and so behaviors can be changed through interactions. Agents can learn and adapt through agent-agent interactions, and agent-environment interactions often related to a specific location and/or sets of conditions, and their interactions can change conditions through feedback loops in which the action of people can shape the environment, which in turn can shape human behavior. The historical exploitation of the Galapagos fisheries has implications for the present as well as the future, as the sea cucumber fishery has not yet fully recovered, preventing the opening of seasonal fisheries and reducing quotas in open years. As communities become more powerful, grow the urban infrastructure, and respond to the needs of residents, they have become more central across the Galapagos Islands. In addition, tourists, attracted to the Galapagos by the environment and the iconic species that reside there, have increasingly sophisticated and "tastes" for costly goods, products, and services. This is an important scale to reconceptualize the Galapagos Islands as they increase their capacity to serve the needs of local residents and tourists and to extend their reach and impact beyond their geographic domains.

A Transition *by* the Galapagos?

From the preceding overview of the transitions that have affected the Galapagos, a picture emerges of the archipelago as the scene of, rather than an actor in these transitions. The most recent episode of transformative change however, still very much in progress, is creating the capacity of the islands need to become their own agent of change. If cities are "transition machines" as Loorbach and Shiroyama

(2016) suggest, then the ongoing urbanization may provide an opportunity despite the current sustainability and liveability issues it has created.

Two notions from complexity thinking may help to frame just how this could work or what it would entail: self-organization and coevolution.

Self-organization

By self-organization we mean organization that is not induced by some central agent inside or outside of the system. This precludes most exclusively government-led initiatives and most "blueprint" or "roadmap" approaches. Self-organization often takes the form of distributed organization. Among the potential benefits of change from the bottom up, as it were, are that the resulting initiatives or structures are often considered to be more resilient and adaptable, precisely because of their distributed nature. Another potential benefit is that, since the initiatives that are part of the self-organization are *local*, they may be more sensitive to the local peculiarities, providing a better fit with the environment.

There are several challenges also, when we want to foster self-organization. For example, the environment is not always favorable for self-organization. Regulation may be in the way, and limited resources can thwart initiatives before they even start. Another major challenge is how to ensure that several local initiatives connect so that they can amount to actual systemic change.[3] Moreover, how can we be sure that the change effected by self-organization is toward a more sustainable and liveable Galapagos? Despite the more likely local fit and potential for resilience and adaptability, there is really no *intrinsic* reason to suppose the outcome will be more sustainable.

Coevolution

By coevolution we mean specifically how the islands' ecosystems and the social systems develop interdependently. From a strict conservationist point of view, this may sound undesirable as it lets go of the idea that the Galapagos can be kept in, or brought back into, a "pristine" state, a state untouched by humans, an example of the accomplishments of natural selection frozen in time. At present, the natural and human systems are already entangled. The natural system – in fact its very status as an ecological marvel – is at the basis of the socioeconomic system and its current relative prosperity. Conversely, this prosperity keeps incentives in place to curb environmental damage and maintain the protected status. Even the ecological systems have evolved to, and adapted to, some of the pressures of the human presence,

[3] One can of course argue that it is not a true case of self-organization of this does not happen, but the challenge remains.

to a point where rolling these pressures back would be damaging. Trueman et al. (2010) assess the current status of plant invasion in the Galapagos Islands and predict the likelihood of future naturalizations and invasions from the existing flora as well as measures to help limit future invasions. Naturalization involves processes by which non-native species spread into the wild and its reproduction is sufficient to maintain its population, even becoming invasive over time. In the Galapagos, approximately 870 alien plant species exist, 34 percent have naturalized, 16 percent are invasive, and 3 percent are transformers, thereby altering habitats and ecological processes. While the Galapagos are at a relatively early state of plant invasion, Trueman et al. (2010) predict that more species from the existing alien flor will naturalize and invade as propagule pressure increases as the human dimension expands through immigration associated with the linked growth between tourists and residents, thereby increasing the services demanded in communities with their potential reach occurring throughout the archipelago and well beyond the islands.

Here, as for self-organization, the issue is not so much to accomplish the phenomenon itself; as argued above, coevolution is already happening, and the challenge is to ensure the dynamics is beneficial for both interrelated systems.

Potential for Self-Organization and Coevolution in Galapagos

There are many examples of emergent processes in Galapagos. Some of which come from fishing, and others are found in tourism. Many of the strategies and practices in the case of fishing were developed by fishermen to increase their production, and in the long term, they created a challenge to the sustainability of the islands. Some of the examples include the introduction of new fishing techniques such long lines, fiberglass boats, nylon nets and lines, outboard engines, and GPS equipment. And the Park has recently instituted fish aggregation devices (FADs) to consolidate fish population for local fishers. These different introductions and inventions at times produced unsustainable practices, but also some of them included more sustainable ideas that if fostered could enhance the management of the commons. Some fishermen claim that they were the first ones to ask for limits to the amount of catches of lobsters and that they lead in the promotion of the idea of creating a marine reserve and the ideas of experiential artisanal fishing (Pesca Artesanal Vivencial). Often many of these activities are later regulated, controlled, and, in some cases, criminalized by the government institutions such as the Galapagos National Park. In some cases this process of regulation of the emergent activities results in the creation of new more sustainable practices, but in others regulation has resulted in the disappearance or the transformation in not very sustainable ways of these potentially sustainable activities. Thus, in the case of day tours and sportfishing, the regulations imposed by the Galapagos National Park of this activity resulted in the creation of sportfishing catch and release or the fishing of smaller fish in an artisanal manner. Larger boats with ever more powerful engines were imposed on the fishermen to accommodate the ever more demanding tourists. Thus, in this process small-scale, slow pace type of tourism was sacrificed for a type of tourism that demanded more comfort and speed. The growth of land-based tourism in the last few decades has resulted in an effort to increase the security and comfort of the tourist but has also resulted in boats having bigger engines that go faster and threaten the local fauna such as sea lions and turtles.

These emergent processes create at once both competition and cooperation between actors. The new emergent processes are also regulated and deregulated. In any case, the systems and organizations created can be more or less sustainable. Emergent processes are often not sustainable from an ecological and social point of view. This is the case of sea cucumber fisheries that resulted not only in ecological problems but also in social and political conflicts. Around the year 2003 as overfishing caused the collapse of the fisheries, there were also different social and economic problems.

Harnessing Complexity for a Transition: Transition Management

As suggested above, the great challenge of a pursuing a transition "from the bottom up" is creating the circumstances in which various small-scale contributions line up, or scale up, and create desirable outcomes, in addition to the challenges faced by the various small-scale initiatives themselves. In practice this may mean that a component of top-down coordination may help to enable the necessary bottom-up dynamics. Transition management (e.g., Loorbach 2010; Rotmans and Loorbach 2009) is a governance approach – or rather a *meta*-governance approach – based on exactly the kind of complexity principles we discussed in this paper. It recognizes that change agents can be found across societal strata and hierarchies and that one of the key challenges is to connect long-term visions with short-term actions. The following draws heavily from Loorbach's (2010) synthesis article.

Transition management observes activities of different "types" (Loorbach 2010) with different foci and time horizons that are necessary to achieve broad societal change:

- Strategic activities such as vision development and long-term collective goal and norm setting. These activities have a long-time horizon (30 years) and relate to the "culture(s)" of the societal system as a whole.
- Tactical activities with a horizon of about 5–15 years dealing with the "structures" of the system. These are the activities that are the concern of strategic planning (which could lead to some terminological confusion).
- Operational activities with a short-time horizon of up to 5 years, like (societal) innovation projects. These activities deal with "practices."
- Reflexive activities, which do not have a particular time scale. These relate to monitoring, evaluation, etc.

While these "types" can be regarded as just an analytical framework to organize observed, e.g., historical, change processes, transition management also proposes specific policy instruments ("tools") to purposefully bring about long-term, structural change as well as harnessing the innovative potential that is already present in many societal systems.

At the strategic level, the key instrument is the transition *arena*. An arena is composed of a small group (e.g., 15 people) of "frontrunners" who are essentially societal innovators. A frontrunner may have an innovative vision, be an entrepreneur making a new technology available, an intellectual or a neighborhood champion. At any rate, the transition arena is not meant to be a forum or representative stakeholders. The arena develops a transition vision and a shared systemic perspective on the issue at hand. Though this perspective needs to be shared to some extent, this does not mean the arena needs to accomplish consensus. Rather dissent is encouraged in order to foster creativity. For an arena to be effective, it is crucial that the participants are well connected and have some influence or authority, though this need not be translated with having a formal power position.

At the tactical level, the transition vision is translated by the arena into a transition *agenda*. This could entail preparing scenarios and, importantly, the identification of – possibly many – pathways. Here also it is not necessary to agree on a single pathway. The operational level has a strong focus on transition *experiments*, which can be seen as "regular" societal innovation projects but with an explicit connection to the societal challenge identified by the arena at the strategic level (e.g., see van den Bosch and Rotmans 2008).

Though the hierarchy of strategic, tactical, and operational suggests a sequence to the transition management activities, this is not at all a requirement, and it is usually presented as a cyclical process (the "Wheel of Loorbach," see Fig. 4.1).

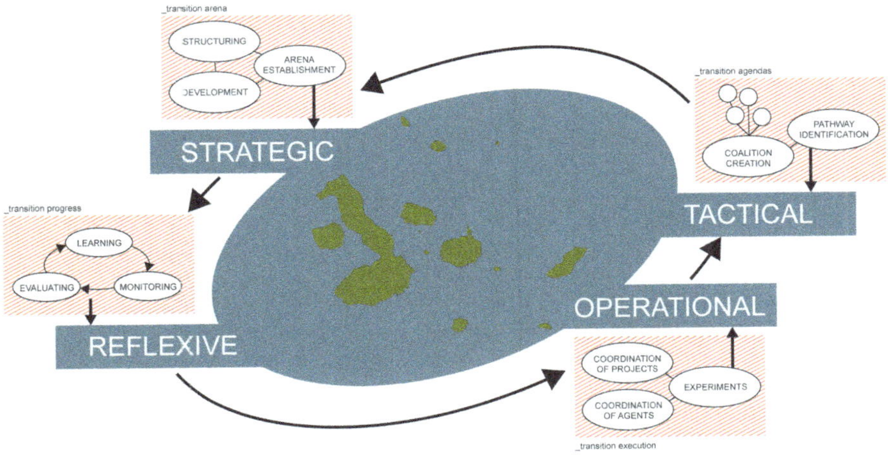

Fig. 4.1 The "Wheel of Loorbach." (After Loorbach 2010)

For example in a case presented by Loorbach and Rotmans (2010), the starting point was a preexisting portfolio of innovation experiments in the Dutch healthcare sector, which were reframed as transition experiments.

Managing the Galapagos Transition?

While it is not the aim of this chapter to provide policy advice or set out guidelines, we do suggest that there is a potential *harmonizing* role to play for a process approach like Transition Management. We sketched a picture of the Galapagos as first being the scene of transitions inflicted by outside influences and recently also by forces from within. Presently, the islands are experiencing much innovative dynamics related with the increasing urbanization and expanding tourism industry. These dynamics also entail various tensions because of conflicting interests, power differences, and the various levels of governance involved. Interestingly, all parties have a definite interest in preserving the Galapagos as an ecological marvel – whether as a unique selling point or out of a sense of stewardship.

There is therefore scope for a transition management process to channel the dynamics and to make creative use of the existing conflicts. A Galapagos transition arena would bring together frontrunners from the communities, conservationists, and government while – within that environment – eliminating the hierarchies and de-politicizing issues. The transition management process and the arena would exist as a process of "shadow governance" and provide continuity beyond political cycles. Of course, one should not expect miracles from transition management – real issues remain to be solved, and this will often result in some parties feeling worse off than others. What transition management does positively add to such situations is a process in which matters in complex systems can be constructively addressed with all the relevant – possibly unexpected – perspectives contributing.

References

An L (2012) Modeling human decisions in coupled human and natural systems: review of agent-based models. Ecol Model 229(25):25–36

Benenson I (1999) Modeling population dynamics in the city: from a regional to a multi-agent approach. Discret Dyn Nat Soc 3:149–170

Brown DG, Robinson DT, Qi J, An L, Rand W (2008) Exurbia from the bottom-up: confronting empirical challenges to characterizing a complex system. Geoforum 39:805–818

de Haan FJ (2018) Place in transitions – concepts for when it matters: essentially, accidentally, locus and nexus. In: Urban sustainability transitions: Australian cases – international perspectives. Springer Nature, Singapore

de Haan FJ, Rotmans J (2018) A proposed theoretical framework for actors in transformative change. Technol Forecast Soc Change 128:275–286. https://doi.org/10.1016/j.techfore.2017.12.017

Entwisle B, Malanson GP, Rindfuss RR, Walsh SJ (2008) An agent-based model of household dynamics and land use change. J Land Use Sci 3:73–93

Gonzalez JA, Montes C, Rodriguez J, Tapia W (2008) Rethinking the Galapagos Islands as a complex social-ecological system: implications for conservation and management. Ecol Soc 13(2):13. http://www.ecologyandsociety.org/vol13/iss2/art13/

Hodson M, Marvin S (2010) Can cities shape socio-technical transitions and how would we know if they were? Res Policy 39(4):477–485

Holling CS (2001) Understanding the complexity of economic, ecological, and social systems. Ecosystems 4(5):390–405

INEC (2010) Ecuador population and housing census 2010. National Institute of Statistics and Censuses (Ecuador), Quito

Kerr SA (2005) What is small island sustainable development about? Ocean Coast Manag 48(7–8):503–524

Kerr S, Cardenas S, Hendy J (2004) Migration and the environment in the Galapagos: an analysis of economic and policy incentives driving migration, potential impacts from migration control, and potential policies to reduce migration pressure. Motu working paper series, no. 03-17. https://ssrn.com/abstract=512062

Loorbach D (2010) Transition management for sustainable development: a prescriptive, complexity-based governance framework. Governance 23(1):161–183

Loorbach D, Rotmans J (2010) The practice of transition management: examples and lessons from four distinct cases. Futures 42(3):237–246

Loorbach D, Shiroyama H (2016) The challenge of sustainable urban development and transforming cities. In: Loorbach D, Wittmayer MJ, Shiroyama H, Fujino J, Mizuguchi S (eds) Governance of urban sustainability transitions: European and Asian experiences, Theory and practice of urban sustainability transitions. Springer Japan, Tokyo, pp 3–12. https://doi.org/10.1007/978-4-431-55426-4_1

Loorbach D, Wittmayer J, Shiroyama H, Fujino J, Mizuguchi S (2016) Governance of urban sustainability transitions: European and Asian experiences. Springer Japan, Tokyo

Malanson GP, Walsh SJ (2015) Agent-based models: individuals interacting in space. Appl Geogr 56:95–98

Pizzitutti F, Walsh SJ, Rindfuss RR, Gunter R, Quiroga D, Tippett R, Mena CF (2017) Scenario planning for tourism management: a participatory and system dynamics model applied to the Galapagos Islands of Ecuador. J Sustain Tour 25:1117–1137. https://doi.org/10.1080/09669582.2016.1257011

Roorda C, Wittmayer J, Henneman P, van Steenbergen F, Frantzeskaki N, Loorbach D (2014) Transition management in the urban context: guidance manual. Dutch Research Institute for Transitions (DRIFT), Erasmus University Rotterdam, Rotterdam

Rotmans J, Loorbach D (2009) Complexity and transition management. J Ind Ecol 13(2):184–196

Schot J, Kanger L, Verbong G (2016) The roles of users in shaping transitions to new energy systems. Nat Energy 1:16054

Sørensen E, Torfing J (eds) (2007) Theories of democratic network governance. Palgrave Macmillan, Hampshire/New York

Torrens PM (2012) Moving agent pedestrians through space and time. Ann Assoc Am Geogr 102:35–66

Trueman M, Atkinson R, Guezon A, Wurm P (2010) Residence time and human-mediated propagule pressure at work in the alien flora of Galapagos. Biol Invasions 12:3949–3960

van den Bosch S Rotmans J (2008) Deepening, broadening and scaling up: a framework for steering transition experiments. Knowledge Centre for Sustainable System Innovations and Transitions (KCT) Delft/Rotterdam, Netherlands

Villacis B, Carrillo D (2013) The socioeconomic paradox of Galapagos. In: Walsh SJ, Mena CF (eds) Science and conservation in the Galapagos islands: frameworks & perspectives. Springer New York, New York, pp 69–85. https://doi.org/10.1007/978-1-4614-5794-7_4

Walsh SJ, Mena CF (2016) Interactions of social, terrestrial, and marine sub-systems in the Galapagos Islands, Ecuador. Proc Natl Acad Sci. http://www.pnas.org/content/early/2016/10/05/1604990113.abstract

Walsh SJ, McCleary AL, Heumann BW, Brewington L, Raczkowski EJ, Mena CF (2010) Community expansion and infrastructure development: implications for human health and environmental quality in the Galapagos Islands of Ecuador. J Lat Am Geogr 9(3):137–159

Walsh SJ, Malanson GP, Entwisle B, Rindfuss RR, Mucha PJ, Heumann BW, McDaniel PM, Frizzelle BG, Verdery AM, Williams N, Xiaozheng Y, Ding D (2013) Design of an agent-based model to examine population-migration interactions in Nang Rong district, Thailand. Appl Geogr 39:183–198

Part II
Socio-ecological Models

Chapter 5
Demographics of Change: Modeling the Transition of Fishers to Tourism in the Galapagos Islands

Stephen J. Walsh, Kim Engie, Philip H. Page, and Brian G. Frizzelle

Introduction

Worldwide, "ecotourism" has been held up as a promised economic engine for local communities living in areas of high biological diversity, proposing environmentally "responsible" travel that also channels profits into conservation efforts and the local community (TIES 2016; TNC 2017). In many cases, however, nature-based tourism has fallen short of such promises, creating as many conflicts as it has intended to solve (Epler 2007; Honey 2008). The Galapagos Islands of Ecuador are a perfect microcosm of these global issues, being a UNESCO World Heritage Site, National Park, and Marine Reserve that is also straining from an explosion of tourism growth since the 1970s, provoking national and international concerns (Epler 2007; Watkins and Cruz 2007; Nash 2009; Quiroga 2009; Pizzitutti et al. 2014, 2016; Walsh and Mena 2016).

The tourism industry is the dominating force in the Galapagos Islands, directly employing 60% of the residents and representing almost the entire economy. The tourism industry is central in determining the dynamics of change on the islands, via the processes of generating high flows of human migration from mainland South America and from international locations, increasing the introduction of invasive

S. J. Walsh (✉) · K. Engie
Department of Geography, Center for Galapagos Studies, University of North Carolina at Chapel Hill, Chapel Hill, NC, USA
e-mail: swalsh@email.unc.edu

P. H. Page
Center for Galapagos Studies, University of North Carolina at Chapel Hill, Chapel Hill, NC, USA

B. G. Frizzelle
Carolina Population Center, University of North Carolina at Chapel Hill, Chapel Hill, NC, USA

T. Kvan, J. Karakiewicz (eds.), *Urban Galapagos*, Social and Ecological Interactions in the Galapagos Islands, https://doi.org/10.1007/978-3-319-99534-2_5

species, stressing ecosystem goods and services, accelerating the consumption of resources, and increasing the pressure on basic services in local communities (Kerr et al. 2004). In parallel with tourism increases has been the growth of permanent residents (~30,000 today), which has expanded in part to provide industry services concentrated in the local communities (Nuñez 2007; Epler 2007).

Rozzi et al. (2010) use the term "mass nature tourism" to describe many practices with "minimal environmental reforms" and "nominal relations" with local people. Over time, the lion's share of Galapagos tourism profits has concentrated in the hands of outside companies. At present, vocal debates over the future of tourism have brought to light several different possible future trajectories of Galapagos tourism development (World Wildlife Fund 2003; Gonzalez et al. 2008; Pizzitutti et al. 2014; Walsh and Mena 2016). If residents retain control of business opportunities and high-paying tourism jobs, the industry may become the envisioned bridge to a robust and satisfying livelihood with lower environmental impacts for many more residents. If most of the market, however, continues to be dominated by outside companies and residents also fail to gain the skills needed for high-level tourism jobs (e.g., multilingualism), then tourism may continue to be seen as a root of social inequalities.

Today, boat- and land-based tourism support the tourism industry as boat berths and hotel rooms provide travel and lodging alternatives to residents and tourists who seek to explore some of the approximately 150 visitation sites that are distributed throughout the archipelago, more or less equally between marine and land access. Boat- and land-based tourism place differential demands on the environment as well as on the local support infrastructure. Historically, boat-based tourism was the principal way to explore the archipelago, with visitors sleeping and dining onboard and viewing the iconic species and iconic settings by disembarking at designated tourist destinations at prescribed days and times. Supported by onboard guides, boat-based tourism flourished, but at the expense of the local population sharing in the generated revenues, as most boat owners were from outside the Galapagos and several from outside of Ecuador. With a boom in the construction of hotels, shops, restaurants, and related services to provide visitors a travel alternative, further enhanced through "island hopping" to nearby islands on locally owned and operated speed boats, local people became more directly engaged in the tourist industry; however, the community infrastructure was further strained, particularly, when additional Ecuadorians immigrated from the mainland to the Galapagos for work in the burgeoning tourism industry (Kerr 2005; Walsh and Mena 2013b; Villacis and Carrillo 2013). In 2001, 30% of the Ecuadorian and 86% of the foreign tourists stayed on boats compared to 9% and 66%, respectively, in 2010. It is typical that boat passengers spend less time on land and are less likely to impact community infrastructure, such as wastewater treatment and energy consumption; however, the environment is at risk of accidental oil spills from boats as well as the accidental and planned illegal dumping of refuse into fragile nearshore areas. Boat-based tourism can also negatively impact iconic species, for instance, through propeller strikes of sea turtles and sea lions. In the Galapagos, the percent of tourists staying on boat has been in decline for several years.

This chapter focuses on one segment of the Galapagos population and their abilities to obtain jobs in tourism – local fishers. Small-scale fishers face very strong forces of change, namely, the growing difficulty of earning a living after decades of declining harvests, low seafood prices, and tensions with managers (Stacey and Fuks 2007; Hearn 2008; Engie 2015). A 2012 survey of 167 people associated with the fishing industry found that a majority of full-time, active fishers wanted to exit the industry and were pessimistic about its future. Engie (2015) also found that many fishers had already transitioned to a permanent state of long-term occupational diversity, straddling jobs in fishing with tourism, government, and local retail. In theory, the overwhelming growth of tourism provides opportunities beyond fishing; however, many have expressed frustration at a perceived lack of participation in the rich tourism economy (Stacey and Fuks 2007). While this desire to participate is strong, the reality is also that the fishing population is unevenly educated, and many people who would like to stop fishing have limited options. Fishers vary in education, job preferences, family needs and household size, and household income.

In this study, we model hypothetical scenarios of tourism development through the *Galapagos fisheries agent-based model* (***GF-ABM***). The ***GF-ABM*** has several parameters that represent key processes related to demographic change and fisher agents skills, for instance, the number and distribution of jobs in fisheries, tourism, and government, the likelihood that the sea cucumber and lobster fisheries are open in any given year, fisher and fisher household expenses, fisher characteristics and skill levels, the influence of the household within the community, and checks on several fisher and household conditions, such as, sailing certification, job preferences, new births/deaths, cost of living, household expenses, and wealth. The key difference in our modeled scenarios is the number and type of local jobs that are created. We focus on the process of local job creation and the combinations of factors that could be most amenable to maximizing the ease of livelihood transitions into tourism for local fishermen. The model outcomes are interpreted relative to changes in tourism and urban structure in the Galapagos Islands, using demographic projections to establish trends and trajectories of change in tourism and the associated residential population. Further, the model attempts a careful assessment of the material benefits that local residents receive through tourism, by tracing job creation in tourism to individual household strategies and wealth. Because both the desire to benefit more from tourism and varying individual skill sets for tourism jobs are widespread among Galapagos residents beyond fishermen, this paper's implications are relevant to Galapagos society in general and not just to fishers.

We highlight the creation of tourism jobs primarily in urban settings, particularly, where tourism services are most prevalent and local people are most engaged. While local people are involved in boat-based tourism, the number of people engaged in land-based tourism is more pronounced. The skill set demanded of some tourism jobs may limit local people, particularly, fishers who may seek to diversity individual and/or household livelihoods by transitioning to tourism, part- or full-time, to counteract uncertainties associated with social and ecological forces of change to local fisheries, for instance, from established fishing quotas and fishing seasons as well as market prices and ENSO events that diminish marine productivity levels and reduce local fish populations.

We use a "complex systems" approach and agent-based models (ABM) to integrate exogenous and endogenous forces that combine to alter social-ecological systems in complex and nonlinear ways (Walsh and Mena 2013a). The ***GF-ABM*** examines the sustainability of a coupled natural-human system in an ecologically fragile area to explore alternate strategies that allow economic development and resource conservation to occur in sustainable ways. The research is framed within "biocomplexity" that encompasses the complex interactions within and among ecological systems, the physical systems on which they depend, and the human systems with which they interact (Manson 2001; Michener et al. 2003; Mahon et al. 2008, Malanson and Walsh 2015, Walsh and Mena 2016).

In our modeled scenarios, we alter the number and type of local jobs created with ongoing tourism. We posit that different arcs of tourism development can create drastically different numbers and types of local jobs, a critical link that has been understudied in the Galapagos. This link also critically impacts the urban and community structure of the Galapagos. Acknowledging that both the desire to work in tourism and varying individual skill sets for tourism jobs are widespread among Galapagos residents, we leverage detailed information on demographics, job preferences, and skill sets among a subset of the Galapagos population, local fishers, collected by Engie (2015). This type of individual-level information is a critical input to labor movements and has thus far been absent from models on Galapagos development.

Study Area

The Galapagos Islands, a National Park, Marine Reserve, and a UNESCO World Heritage Site, are world renowned for their iconic species, iconic landscapes, and iconic marine environments. Part of Ecuador and located in the Eastern Pacific Ocean 1,000-km from the continent, the Galapagos archipelago contains 11 large islands and 100s of smaller islands and islets that contain some of the world's most unique plants and animals (Fig. 5.1). With high endemism, the Galapagos Islands support some of the most recognizable iconic species on the planet – giant tortoises, marine iguanas, Darwin finches, and so much more. In recognition of this biological diversity, the Galapagos National Park (GNP) was established in 1959 by the Ecuadorian government with the help of IUCN and UNESCO, covering 97% of all terrestrial land in the archipelago. The Park was inscripted as a UNESCO World Heritage Site in 1978, which expanded to include the Galapagos Marine Reserve in 2001 (Rozzi et al. 2010; UNESCO 2010).

While its endemic flora and fauna evolved completely cut off from human settlement, the Galapagos Islands are now being transformed by expanding human settlement. With 30,000 residents living mostly in coastal tourism communities on four populated islands, the Galapagos Islands in 2016 welcomed approximately 225,000 visitors to see and explore the archipelago as boat- or land-based tourists, with the

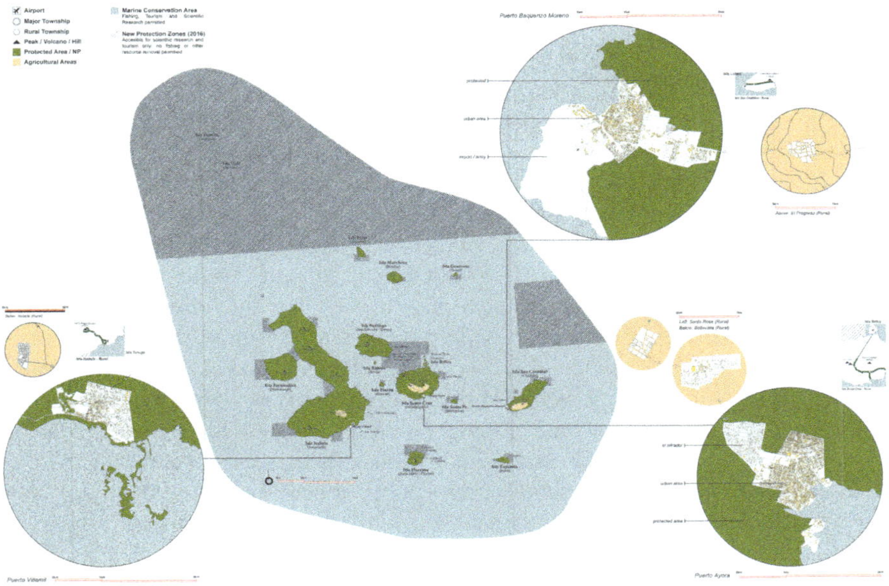

Fig. 5.1 Main islands of the Galapagos Archipelago of Ecuador

central goal of interacting with highly accessible and unique iconic plant and animal species that define these "Enchanted Islands."

With more and more people living and visiting the Galapagos Islands, more and more products and materials are being imported into the islands on cargo ships (and airplanes) that are in perpetual movement between the Galapagos and the continent, bringing all the necessary products to support the expanding human dimension, including the planful and accidental introduction of alien species. Often outcompeting native and endemic plant species, invasive species are capable of transforming landscapes and diminishing land productivity, often contributing to field and farm abandonment, particularly, when job opportunities in tourism, in concert with employment preferences and skill levels, encourage household employment diversification and livelihood transition from farm to community living, further stressing a deficient urban infrastructure for residents and tourists. Mindful of these challenges to the sustainability of island ecosystems, the Galapagos National Park, together with several government and nongovernment organizations, work toward the eradication of introduced and invasive species, but, so too, the management of people and their necessities in the Galapagos, generally consolidated in communities, whose infrastructure has languished behind the "peopling of the Galapagos," thereby, creating challenges to the island's sustainability and ecological vitality (Percy et al. 2016). Now explicitly linked through a coupled human-natural system perspective that recognizes the complex interplay of the social, terrestrial, and marine sub-systems, the Galapagos Islands are best understood through an integrative approach that addresses the challenges and implications of human-environment interactions.

Alternate Household Livelihood Strategies

How a household makes a living through the combination of assets, capabilities, and activities is encompassed in the notion of a *livelihood* (Chambers and Conway 1992; Ellis 1998, 2000). More than the types of occupations people have, a livelihood is about the relationship between a household and its social and natural environment (Qin 2010). A livelihood includes rules that govern access to natural resources, and coping strategies, such as labor diversification that helps minimize subsistence risk in the context of environmental change (Haglund et al. 2011). Further, a livelihood is a dynamic set of activities that may change seasonally according to available opportunities and assets or change from year to year in relation to economic trends, exogenous shocks, and changing social and political institutions (Ellis 1998, 2000). Livelihood diversification focuses on seasonality, labor markets, risk management, and coping behaviors to address the relationships between jobs, labor allocation, and income (Ellis 2000). Seasonal diversification is based on the idea that the seasonality can lead to large intra-annual fluctuations in both income and labor. Thus, households diversify their livelihoods to reduce income instability (Winterhalder et al. 1999). They are a collection of individuals that consume or contribute to annual income and overall wealth as a consequence of demographics, job preferences, job availability, job and skill levels, income levels, educational status, power relationships to community leaders, cost of job switching, status of fisheries (e.g., quota or season), status of tourism (e.g., number of visitors, number of jobs, level of jobs), and status of government (e.g., number of tourists and residents). The goal in the Galapagos is to examine annual changes in household income and accumulated household wealth as a consequence of job diversification relative to the ability to seize employment opportunities by members of the household and to adjust to incentives (e.g., more tourists/more jobs) and disincentives (no lobster or sea cucumber season or a low allowable catch/quota).

Changing Island Economies

Four sectors are important to understanding economic development on the islands today: agriculture, fisheries, tourism, and the public sector. Agrarian practices brought human settlement to the islands, with plantations and farms gradually established from around 1832 onward by Ecuadorian colonists (Ospina 2006). However, farming has been shrinking with migration to urban areas and the advent of tourism in the 1970s (Ospina 2006; Epler 2007) and employed only 8.4% of the labor force in 2009 (INEC 2010). Similarly, artisanal fishing,[1] long an activity engaged in partly for subsistence and as part of diverse livelihoods along with tour-

[1] Only small scale, fishing is allowed in the Galapagos Marine Reserve (GMR), conducted by registered local residents. Commercial fishing was banned with the establishment of the GMR in 2001.

ism and other commerce jobs, has been shrinking since the early 2000s (Engie 2015). A spike in profitability with gross income peaking at $7 million USD in 2003 (Hearn et al. 2005) was short-lived, fueled by a then newly exploited sea cucumber fishery and its rapid depletion (e.g., Hearn 2008). Even in 2007, Watkins and Cruz (2007) estimated fishing as <4% of total Galapagos income.

While now minor economically, livelihood transitions away from fishing are important for several reasons. First, the activity has underlain the greatest tensions between managers and the local community in recent decades, when multiple strikes by fishers shutdown civil institutions and helped fuel a governability crisis in the islands (e.g., Ben-Yami 2001; Stacey and Fuks 2007; Hearn 2008). In addition, the fishing sector has been the focus of particular attention by managers, asked to bear the most direct cost of conservation to individual livelihoods by giving up fishing grounds for protected areas, but promised special future assistance in transitioning to non-fishing livelihoods. While their sea-based skills might seem to translate well into the boat-based tourism industry, many have expressed frustration at a perceived lack of participation in the rich tourism economy (Stacey and Fuks 2007; Engie 2015). Therefore, helping fishermen transition into tourism is a long-standing issue with strong local political salience.

Galapagos communities have transitioned from livelihoods based on farming and fishing to an economy dominated by tourism-related employment and a robust governmental public service sector. From a trickle of visitors in the 1960s, tourism has exploded into nearly 225,000 tourist visitors in 2016. While exact numbers are hard to find, it is clear that tourism now drives economic development on the islands. reported 40% employed in tourism in the late 1990s. In 2002, Epler (2007) similarly noted tourism as one of the largest employment sectors, pegging 22% directly employed, and far more indirectly.[2] In 2009, those in transport, hotels and restaurants, and commerce combined made up 43.7% (INEC 2010), many likely linked to tourism. Finally, as a semiautonomous Ecuadorian province and one that makes important contributions to the central government in Quito from tourism revenues, the municipal and provincial levels of government are important employers on the islands. In 2009, 12.8% were employed in public administration provincially, rising to almost a one-quarter (22.8%) on San Cristobal, the seat of the provincial government (INEC 2010).

Nature-Based Tourism and Local Employment

Despite efforts to purposefully structure tourism policies to benefit the local community, the balance of profits has slowly skewed to non-Galapagos residents and outside companies and raised ire in local communities over time (i.e., Ospina 2006; UCOOPEPGAL 2012). The unanticipated popularity of the Galapagos Islands as a

[2] Epler counted 1,929 people directly employed in tourism from various surveys out of 8,772 people economically active as of 2002.

tourism destination, with revenues "…exceeding anyone's wildest expectation" (Epler 2007, iii), has subjected many ongoing policies to profit seeking from outside interests, at times at the expense of the intended beneficiaries of local tourism opportunities, Galapagos residents. One illustration is the "cupo" system of licenses for cruise operators that began in the 1980s, which is a quota system specifying how many passengers each cruise operator can carry (Epler 2007). Cupos cannot be legally sold and were originally given to local residents. However, Epler (2007) and both report a buying up of cupos by larger companies from small operators, and ways of the transfer of ownership occur, such as paying monthly rents to local owners or making agreements transferring rights of operation, without the local owner involved in any business dealings. With a conservative yearly estimate of $1.35 million USD revenue for a 16-passenger cupo (Epler 2007), outside companies can end up reaping the lion's share of profits compared to local owners. Similarly, a newly created license for "experiential fishing" aimed solely at local fishermen to give them a livelihood in tourism, has been subject to legally dubious agreements of business rights with outsiders, where a local fisherman makes a fraction of profits and is not the true owner of a sport fishing operation (Schuhbauer and Koch 2013; Engie and Quiroga 2014).

Adding to the complexities of moving toward a tourism model that is more beneficial to local communities, while the desire for greater local control of tourism businesses is strong, the reality is that Galapagos residents have uneven capacities and skill levels for various types of touristic livelihoods. Galapagos tourism was purposefully designed as a world class, high-end destination where visitors could expect the highest level of service and care from tour operators (Epler 2007; Grenier 2007). After 1998, only Galapagos residents could be employed as naturalist guides, an occupation once open to all. However, "the average quality of Galapagos guides who, for four decades were considered among the most highly qualified in the world, rapidly decayed…" (Rozzi et al. 2010, 12). Sought-after skills in the tourism industry include the ability to speak English and general sales abilities, often afforded through higher education beyond the minimum 4 years of high school. Engie (2015) found that while fishermen had great seafaring skills, they varied greatly in their educational levels, multilingualism, job preferences, family needs and household size, household income, and basic desire to engage in customer service. While there was a strong desire to participate more in tourism, many who would like to stop fishing had limited options.

The rapid pace of Galapagos development has been noted with alarm for at least a decade (Watkins and Cruz 2007). The wealth and fame that tourism has brought to this iconic archipelago has been a double-edged sword, funding many conservation efforts through visitor and other fees and undoubtedly granting some economic benefits to communities. However, fragile volcanic ecosystems, naturally limited freshwater and arable land, have been strained. The vast majority of food and water consumed are not locally sourced but imported in ever-greater quantities on cargo ships and airplanes from South America, bringing along with them both unintended hitchhikers such as rats, seeds, and associated nonnative pathogens and intended introductions such as ornamental plants and agricultural crops. Often outcompeting native and endemic plant species, invasive species have transformed landscapes and diminished land productivity (e.g., Henderson and Dawson 2009; Nash 2009; McCleary 2013). Rozzi et al. (2010) note that invasive plant species now outnumber native plant species. With plans for new hotels and tour cruises continuing to be proposed, "local tourism" has become a strong political issue, and one whose urban footprint must be considered, since local- and land-based tourism are often synonymous, given the loss of real ownership of boat-based operations that locals have experienced over the years. Future possible trajectories of tourism development are, therefore, crucial to consider. While the future is impossible to predict, multiple possible scenarios of future touristic development can be imagined, including:

a. Anemic job growth for local residents, where existing boats and new hotels proposed across the archipelago in 2017 hire largely immigrants from mainland Ecuador for lower wages.
b. Job growth dominated by demand for unskilled and semiskilled positions in manual labor and as sailing crew, cleaners, truck and boat drivers, and fishing guide assistants. Residents are offered work in tourism largely as hired help and rarely as owners of tour operations.
c. More vertically, even job growth in tourism for local residents across tiers of skill and ownership, where positions are created in manual labor, semiskilled work as described above, and also at the highest skill levels of business owners and operators both in name and practice.

We argue here that the history of tourism policy, and local job growth trends, must inform future considerations of tourism development. Models can be potent tools for policy-thought experiments. While island-level GDP has been modeled (Taylor et al. 2008) to assess potential job availability and uptake in a population of individuals with heterogeneous job skills and preferences, individual-level data on occupational history and preferences are needed to address more nuanced relationships between people and environment. We investigate the aggregate outcome of the processes of tourism development using an agent-based model to synthesize institutional regulatory constraints, hypothetical market conditions, and individual decision-making to scale up to societal trends in nonlinear ways.

Complexity Theory and the Galapagos Islands

An integrative approach that addresses the challenges and implications of human-environment interactions strongly fits research needs focusing on the Galapagos Islands. A coupled human-natural system perspective explicitly recognizes the complex interplay of the social, terrestrial, and marine subsystems in the Galapagos Islands (Miller et al. 2010; Walsh and Mena 2016). Identification of critical feedback of coupled natural-human systems is fundamental to understanding population-environment interactions and the plausible scenarios that perturb system dynamics and alter system behaviors. Our model integrates social-ecological relationships that represent the behaviors, dynamics, perturbations, self-organizations, emergences, and feedback that affect coupled natural-human systems and their transitions. We conduct computer-based "experiments" with a modeling environment to evaluate the degree of codetermination of outcomes in the human (e.g., human demography and household livelihoods/diversification strategies) and ecological (e.g., land use/land cover change and patterns of land suitability) systems. Our goal is to better understand the implications of alternative resource conservation and economic development scenarios in an internationally important, protected area, with direct attention to job diversification of households through alternate strategies that either retain fisheries in fishing as their primary economic activity or facilitate their departure from full-time fishing to part-time fishing or permanent or temporary transition to tourism and/or government employment in the Galapagos Islands as an alternative household livelihood strategy.

The ***GF-ABM*** was developed to examine human-environment interactions through scenarios of change in agent behavior, particularly, of fishers and their households as a consequence of the "pushes" from fishing and the "pulls" into tourism and/or government that recognizes the social and ecological forces that influence job diversification as part of an alternate household livelihood strategy. The ABM modeling environment and the use of Complexity Theory to guide the development of the ***GF-ABM*** relies on several modeling advantages to achieve our goals:

(1) Many households can produce alternative responses to the same problem, based, in part, on their environmental settings and, in part, on their socioeconomic characteristics.
(2) Experiments provide insights into what processes are involved, their scale dependence, and links to other processes and to the generation of emergence.
(3) ABMs develop variability rather than reduce it.
(4) Self-organization can limit divergence of model outcomes over short periods, but over longer periods, exogenous forces may overwhelm any self-organization.
(5) Heterogeneous actors, described through their beliefs, preferences, circumstances, social networks, and behaviors, provide the opportunity for a more nuanced examination of pattern-process relationships.
(6) Pattern can be used to infer process, and process can be used to understand pattern.

Beyond the advances in theoretical frameworks and modeling approaches, our model has important practical uses, including the generation of linkages among variables that were represented in disparate and fragmented date sets. More importantly, our models have been used to illustrate the alternate outcomes associated with different scenarios of human-environment interactions, resulting from the motivations and decision-making authority of politicians and resource manager's that have directly contributed to conservation of a fragile place.

Methods

Under current conditions, Galapagos, being a nature tourism destination that attracts thousands of tourists each year, is near its saturation level. If tourism continues to grow at the pace of the past two decades, depending on the type of growth, in a few years, there will be no open space to ensure that every tourist visiting the Galapagos experiences nature tourism at a high standard of quality. In parallel, social, administrative, and organizational systems in the Galapagos have not demonstrated the ability to evolve at the same rate that has seen the expansion of tourism, creating imbalances and failures as well as interactions that are negatively associated with the tourism sector and amplified for fragile and sensitive island ecosystems. The below model acknowledges the current conditions in the Galapagos and how social-ecological dynamics shape human behavior as well as environmental trajectories of change.

The Galapagos Fishers-Agent Based Model (GF-ABM)

Our model, in a three-part causal chain, (1) looks to household assets, attributes, and responsibilities and (2) asks how they interact with individual preferences, major policy shifts, and environmental fluctuations to (3) influence livelihood strategies. We identify a small set of hypothetical future scenarios, controlled by a set of key system interactions. These identified scenarios are modeled to assess their effect on human well-being via income effects, labor flows among occupations, and the socioeconomic distribution of households in obtaining tourism jobs. Maximal benefit to local Galapagos residents will be based on the rate of job switching into tourism, at any level.

The model's data and design are based on survey data of fisher livelihoods conducted by Engie (2015), which recorded the career trajectories, household capitals, social responsibilities, and job preferences of former and active fishers (N = 166). This survey was administered from March to August 2012 on the island of San Cristobal, the site of the birthplace of commercial artisanal fishing in the Galapagos Islands. Questions drew upon past work on the social aspects of fishing livelihoods and job switching (e.g., Scoones 1998; Terkla et al. 1988; Ellis 2000; Pollnac et al. 2001; Cinner et al. 2009). More methodological details can be found in Engie (2015). While there are a handful of women who are actively fishing in the Galapagos fleet, all the fishers interviewed in this study who actively went to sea were male.[3] As in Engie (2015), we summarize the various livelihood pathways that individuals have chosen over their careers in three categories: (1) fishing intensification, (2) migration out of fishing, or (3) cross-sector diversification represented by long-term, part-time fishing.

[3] A few females with fishing licenses were interviewed, however, all were boat owners, and none actively fished at sea.

The ***GF-ABM*** examines the transition of people away from fishing and into tourism. It does so by looking at three central elements of job transitions. First, it models out major "pushes" and "pulls" affecting local Galapagos residents when making job decisions. Galapagos fishermen, like many others, have long been compelled to consider job diversification (Ellis 2000; Wilen et al. 2000; Ospina 2006) to confront the socioeconomic uncertainty and changes in fish concentrations due to fluctuating marine productivity in the highly dynamic waters of the archipelago that are strongly affected by ENSO (El Nino – South Oscillation) events. Policy uncertainty is also high given potential change in fishing quotas, seasonal closures, and license requirements managed by the Galapagos National Park. Secondly, the model touches upon the individual preferences of the job seekers themselves. Not all fishers wish to transition to non-fishing jobs, nor are people equally equipped for this transition. Finally, it recognizes that not all jobs are created equally within Galapagos economic sectors, by assigning levels to tourism and government jobs that correspond to increasingly higher skill requirements and salaries. We create three tiers of levels in each sector, with Level 1 being the least skilled and Level 3 the most skilled.

Fishers are agents, each modeled individually, are organized into households that can also contain non-fishing members. This structure replicates the reality of the varying social supports that fishers have, which affect their income needs, job preferences, and available financial and social capital. The income and expenses calculated for each agent is affected by the structure of their household. Households are also agents and comprise a collection of individuals that collectively consume or contribute to annual income and overall wealth. The ***GF-ABM*** contains parameters relating to (1) demographic change, (2) livelihood skills, (3) available jobs on the islands, (4) the opening of sea cucumber and lobster fisheries, (5) household income, and (6) household expenses (Table 5.1). We have translated the household survey of fishers conducted by Engie (2015) into a mathematical calculus of decision-making and probabilities of job switching based on certain conditions. For modeling purposes, real fisher behavior is stylized here into chains of decisions, all represented as probabilities. Modeling probabilities of various events are also informed by the semi-structured interviews (N = 127) and participant observation aboard fishing vessels conducted by Engie (2015) to contextualize in as broad a manner as possible, decisions around job switching from the perspective of the individual fisherman.

Initial conditions are set to one fisher per household. Fisher agents are created with varying fishing skills and assets; those with a higher skill level are awarded a sailing certificate. If a fisher owns a boat, the boat maintenance variable is set to the corresponding value. *Non-contributing members* are household members that are over 70 years old, who do not earn any income. *Contributing members* (under 70 years old) may continue to reside with the family, and if so, they are part of the working household and can generate family income and contribute to household wealth. *College students* are adults in the household between the ages of 18 and 21, who are in college and earn only a fraction of their potential income, and *children* are household members under the age of 18, who do not earn any income. An *influence* variable has been created as an index of the amount of influence a household has in the community, a proxy for the level of connectedness that a household enjoys with people and organizations of power and authority.

Table 5.1 Parameters of the ***GF-ABM***

Section	Parameters
Demographic change	Birth rate Probability of new household members Non/contributing income
Livelihood skills	Growth in English language skills Growth in maritime skills Minimum investment needed to obtain a professional sailor certification Decision by any new household member to register as a commercial fisher Influence variable of social capital
Growth and distribution of jobs	Total available jobs Job distribution among tourism, fishing, and governmental sectors three subsections in each sector of increasing skill requirements
Availability of sea cucumber and lobster fisheries	Probability of a fishing season opening, per year
Household income	Annual wages per sector Percentage of total jobs distributed to each island
Household expenses	Maritime certifications Annual cost of living Per adult, child (under 18), and college student Cost of switching job sectors or levels within sectors Maintenance for *pangas* and *fibras* (local boats)

GF-ABM Functions and Key Process Characteristics

The model contains a series of functions and procedures that repeat once for every annual iteration. Figure 5.2 shows the central elements of the ***GF-ABM***. Parameters and procedures are briefly described below, following Walsh and Mena (2016).

In the employment component of the ***GF-ABM***, fisher agents select jobs among three employment sectors (i.e., fisheries, tourism, and government). Note that the agriculture sector is not included because the industry is in dramatic decline and represent by far the fewest number of jobs, which are commonly hired on contract from mainland Ecuador. The tourism and government sectors each have three tiers of jobs that require increasing agent skills, particularly, the ability to communicate in English.

Of particular note relating to environmental change is the *Set Season* parameter to determine whether the sea cucumber and/or lobster seasons are open. Also, the *Die-or-Age* procedure performs a "death check" on every fisher agent, using the agent's age, sex, and a derived mortality rate. If the agent dies, the model checks to determine if the fisher owns a boat, and if so, the model calls to the *Transfer-Boat-Ownership* subroutine. The *New Birth Check* procedure performs a birth check on all households, sorted by age and sex, and the *New Household Additions* procedure checks each household for possible new non-birth additions, considered to be rela-

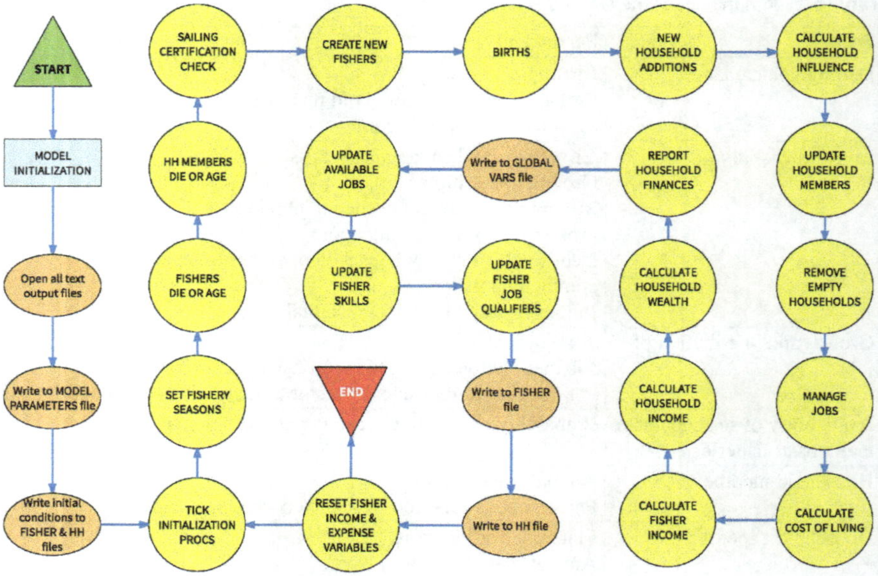

Fig. 5.2 Structural design and key elements of the Galapagos fishers agent-based model (***GF-ABM***)

tives or current household members. The *Update Household Influence* procedure updates after all fishers and non-fishers have died or aged, and new members have joined the household. The *Update Household Members* procedure updates the number of fishers, non-fishers, and total number of members for all households. The *Manage Jobs* procedure allows every fisher agent to select jobs and percent time each year in fishing, tourism, and government. The *Calculate Household Cost of Living* procedure updates the household conditions using a *Calculate Finances* to derive income and expenses for fishers and households based on percent time in fishing, tourism, or government, job level, open season in sea cucumber and/or lobster, and boat ownership and type. The *Maximum Possible Tourism Income* is randomly drawn from an income range linked to the employment sectors and job levels. A *Bonus Income* for an open season of sea cucumber is linked to the ability of fishers to earn additional income in those years as well as additional income (and expenses) through boat ownership. In the case of lobster fishing, most years, the fishery is open, and so income generated is fundamental to fisher livelihoods, depending upon established quotas and market conditions. The *Calculate Household Income* procedure calculates the income from fisher and non-fisher (contributing and non-contributing) household members, including income derived through boat ownership and fishing activities. Lastly, the *Calculate Household Expenses* procedure calculates the total expenses for each household for every year from multiple sources, and the *Calculate Household Wealth* procedure updates wealth for each household for each year, accounting for annual income, less annual expenses.

Power Relationships and *Influence* in the ***GF-ABM*** using the concept of social networks to create links of individuals and households to key actors and families. Core nodes in the social network represents, for instance, ties to the Director of the Galapagos National Park, mayor of the community, governor of the Province, president of the fishing cooperative, and owners and managers of hotels and restaurants, and guides and managers linked to the airline and tourism industries. We select some number of "power links" in the community, 4–5 families representing perceived power families and organizations and then randomize fisher connections over time. *Resident and Tourist Projections* are derived through demographic projections of the expanding human dimension, reflected through population census years, social survey, and focus groups. Population migration is a fundamentally important process for the "peopling" of the Galapagos Islands through tourists (i.e., temporary migrants) and residents, primarily, from the Ecuadorian mainland who have come for jobs, primarily in tourism, and motivated by the "pushes" and "pulls" of the sending and receiving areas.

Memory of Agents, not used in the current model version, although it may be incorporate in a subsequent iteration, is useful to simulate human-environment interactions that change as a consequence of "memorable" events that alter human behavior and the development of alternate household livelihood strategies to confront uncertainty and opportunity. ENSO (El Nino Southern Oscillation) events are particularly significant in the Galapagos Islands as water temperature increases, upwelling is disrupted, and marine productivity decreases, with implications observed throughout the food web and the social systems that rely upon fisheries as an economic alternative. Particularly important are the severe events that may redefined human-environment interactions in the Galapagos Islands, including the major ENSO events that occurred in 1982–1983 and 1997–1998. From 2000 to the present, ENSO events occurred in 2002–2003, 2004–2005, 2006–2007, 2009–2010, and 2014–2015, with intervening neutral periods and La Nina conditions that tend to heighten the recovery from El Nino conditions through a deepening of marine upwelling, increase in marine productivity, and recovery of iconic species. Functions can be derived that identify contemporary fishers who are active in the fishing industry during major and minor ENSO events. Hypotheses can be developed to examine possible mechanisms for developing a propensity for adaptation through knowledge, acquired directly or indirectly, linked to personal traits, such as education, that may exhibit a time-decay function for memory, adaptation, and change.

Alternate Household Livelihood Strategies involve categories of employment that are accounted for in the model, tourism, fisheries, and government (Note that agriculture is the fourth employment sector, but its importance in the Galapagos has ebbed). The capacity to switch among employment sectors, primarily to tourism, is based upon job availabilities, regulatory constraints, individual characteristics, job preferences, and skill levels. While job preferences to transition out of fishing (full or part-time), individual/household income, wealth, and assets are important considerations. For instance, a fisher may choose to switch to tourism during an El Nino event, hire out to work in the highlands in farming, or even out-migrate to the continent and return money to the household through remittances. *Job Switching* is associated with the learning of English as a highly desirable skill for employment in tourism. Education level, experience in tourism, and intention to leave fisheries are important factors in developing English language skills. We derive a function that determines the probability of increasing a skill based on age and education and modify that probability based on their job and level of experience.

Seafood Market Demand is highlight through market demand for lobster as of 2013, when 80–100% of the lobster catch in the Galapagos Marine Reserve entered the international market, thereby assuming that the international price of lobster drives local prices in the Galapagos. Thus, we used the weekly prices of frozen lobster tails posted by the New York Frozen Market (NYFM) as our proxy. The lobsters sold in this market are imported from South and Central America, including Ecuador. Setting the sea cucumber season is determined through park decisions based on the abundance counts generated for marine sampling locations, but the political climate in the Galapagos is also a factor. Other options are linking the market prices and the sea cucumber season to ENSO events.

Model Scenarios

Five different model scenarios were run, representing hypothetical changes in the number of jobs in the fishing, tourism, and public sectors, as well as the skill levels needed for these jobs (Table 5.2).

In each scenario, only minor alterations were made with all other baseline inputs held constant (Table 5.1) to maximize the clarity of impact that hypothetical changes made on individual livelihoods and household wealth. For each scenario, the model was run 100 times.

Note that the distribution of jobs within the ***GF-ABM*** reflects job availability as seen from the perspective of individual fishers, and not the distribution of jobs across the Galapagos economy. The modeled jobs, however, are based on the number of fisher agents at initialization. In all of our model runs, the total number of jobs is equal to 1.5 times the number of fishers or 249 jobs available across the industries for the 166 fishers. We split those jobs into the three industries (fishing, tourism, government) and further divide them by job type (Levels 1–3) based on the parameters listed in Table 5.2. For instance, the baseline distribution of jobs shown in Scenario 1 in Table 5.2 (i.e., fishing 60%, tourism 20%, government 20%) results in 149 fishing jobs, 50 tourism jobs, and 50 government jobs at initialization. This distribution is compatible with tourism likely accounting for at least 50% of all jobs in the Galapagos economy.

Scenario 1 simulates what represents an ideal trajectory, where the number of local jobs available in tourism continues to grow, while the number of people fishing shrinks. The model is set so that as individuals move out of fishing, they are absorbed into tourism and the public sector, with the bulk of fishermen able to move into tourism. Scenario 2 simulates a situation where the number of jobs in tourism remains

Table 5.2 Model scenarios of the ***GF-ABM***

	Baseline model	Scenario tests
Scenario 1: Decline in fishing jobs and an increase in available tourism jobs	Distribution of jobs across sectors: Tourism 20% Fishing 60% Public Sector 20%	Distribution of jobs across sectors: Tourism 50% Fishing 20% Public Sector 30%
Scenario 2: Downgrading of available tourism jobs	Distribution of job types in Tourism: Level 1 60% Level 2 20% Level 3 20% (high skill)	Distribution of job types in tourism: Level 1 50% Level 2 30% Level 3 20% (high skill)
Scenario 3: Downgrading of available public sector jobs	Distribution of job types in public sector: Level 1 60% Level 2 20% Level 3 20% (high skill)	Distribution of job types in public sector: Level 1 50% Level 2 30% Level 3 20% (high skill)
Scenario 4: Shrinking fishing opportunities	Yearly probability of open seasons: Sea cucumber (33%) Lobster (85%)	Yearly probability of open seasons: Sea cucumber (25%) Lobster (25%)
Scenario 5: Job growth in the public sector	Total proportion of jobs in the public sector (20%)	Total proportion of jobs in the public sector (25%)

constant, but a slight upgrading of job skill level occurs over time. The proportion of jobs within the lowest skillset declines, used here to represent people who own and operate businesses, high-end naturalist guides who speak English, and those with high certifications (i.e., captain's license) or salaried positions with tour companies. These tour occupations in general bring in the greatest levels of income in the Galapagos. Similarly, Scenario 3 represents an analogous upgrading of available jobs in the public sector. In Scenario 4, we explore the possible effect of shrinking fishing opportunities in the two most lucrative fisheries, lobster and sea cucumber, in a hypothetical situation where the season openings for these two fisheries were drastically less likely. In such a situation, it is possible that fishing jobs would still be available; however, the yearly incomes earned by many fishermen would become much reduced. Finally, Scenario 5 explores the potential effects of changes to other factors influencing livelihood strategies that of rising costs to switch livelihoods.

Discussion

The development of the ***GF-ABM*** was achieved by reliance upon (1) the data collected as part of a survey of fisheries in 2012, described by Engie (2015), (2) rules and relationships specified in programming code that were generated through empirical observations, survey data results, theoretical constructs, and findings reported in the literature, and (3) model runs and the generation of reasonable, expected, and occasionally, surprising outcomes.

There are several challenges in using spatial simulation models outlined above to enrich process understanding and our ability to assess system behavior. Validation of model processes remains a critical concern, because the underlying processes that influence derived patterns are not fully observable across a range of space-time scales. In the future, additional efforts will be devoted to process models and less about characterizing spatial pattern, although advances in measure technologies will be important to model calibration, validation, and projection. Understanding complex systems may challenge our ability to infuse complex processes into our models, cognizant of the non-stationarity of variables and processes as a possible motivation for the generation of near-term projections, and the imposition of greater uncertainty with time in longer-term model simulations. Creation of system patterns through parameterized models may relate to observed or expected outcomes without necessarily defining the corresponding processes involved and their operation. Pattern may be used to infer process, and process may be used to understand pattern. Model form, level of complexity, and degree of uncertainty may limit policy assessments and its utility by decision-makers. The need exists to create useful measures of uncertainty and to place error bands around variables, representation of processes, and assessment of system behaviors and dynamics.

Reflecting upon our ***GF-ABM***, prior modeling activities in our other study sites, and findings reported in the literature, we note some of the more important implications for ABMs and Complexity Theory more generally:

(1) Additional development should focus on process models and less about characterizing spatial pattern, although advances in measure technologies will be important to model validation and projections.
(2) Understanding complex systems may challenge our ability to infuse complex processes into our models and across space and time scales, cognizant of the non-stationarity of variables and processes as a possible motivation for the generation of near-term projections, and the imposition of greater uncertainty with time in longer-term model simulations.
(3) Feedback mechanisms that link human behavior to environmental patterns and processes are complex, offer several measurement challenges, and maybe be scale dependent.
(4) The intention to predict behavior and to explain processes may be mutually exclusive, depending upon scales, parameters, and intentions.
(5) The creation of system patterns through parameterized models may relate to observed or expected outcomes without necessarily defining the corresponding process involved and its operation.
(6) For dynamically represented processes, very different outcomes may be generated.
(7) Model form, level of complexity, and degree of uncertainty may limit policy assessments and its utility by decision-makers.

(8) Heterogeneous actors, described through their beliefs, preferences, circumstances, social networks, and behaviors, provide the opportunity for a more nuanced examination of pattern-process relationships.

Conclusions

Worldwide, the social-ecological links in areas simultaneously reliant on the integrity of ecological systems and burgeoning tourism industries are poorly studied (Mena et al. 2011; Miller et al. 2014). In the past decade, "ecotourism" has expanded globally, with no area of the globe untouched by human impacts. The Galapagos Islands are on a very short list of systems around the globe that are of extreme priority, all of which are challenged and relatively unexplored in the appropriate and sustainable mix of ecological protection and economic development as they confront the impacts of the expanding human imprint on vulnerable environments. Sustainability is the key objective in all locations, with a focus on the preservation of the key attributes of the social-ecological system.

The primary demographic variable driving the rapid demographic growth in the Galapagos is migration, with the overwhelming majority of migrants coming from the Ecuadorian mainland. While a small number might migrate for work in fisheries or other economic sectors, we assume that the increase in tourism is the engine driving the growth in the number of migrants, who come to work in businesses directly catering to tourists and in business/governmental jobs that have grown to support the increased number of tourists and residents. We further assume that future growth in the resident population will be tied solely to growth in tourism, and not to increases in agriculture, fisheries, manufacturing, or any other economic sector

The ecosystems of the Galapagos Islands have evolved for millions of years, mostly due to geographic isolation and the creation of unique and undisturbed settings. Human presence in the archipelago now implies globalized connectivity, thereby eliminating the insulation and protection induced by space and remoteness. The simulations and projections of this study suggest that levels of environmental quality and social-ecological well-being vary according to the level of local and touristic population growth, availability of jobs in tourism and the public sector, and alternate household livelihood strategies of fishers. In addition, under current conditions, Galapagos, being a nature tourism destination that attracts thousands of tourists each year, may be near its saturation level. If tourism continues to grow at the pace of the past two decades, depending on the type of growth, in a few years, there may be no open space to ensure that every tourist visiting the Galapagos experiences nature tourism at a high standard of quality (Pizzitutti et al. 2016). In parallel, social, administrative, and organizational systems in the Galapagos have not demonstrated the ability to evolve at the same rate that has seen the expansion of tourism, creating imbalances and failures as well as interactions that are negatively associated with the tourism sector and amplified for fragile and sensitive island ecosystems.

Fishers are among those who seek part- or full-time employment in tourism (and government) depending upon skill levels, personal preferences, demographics, job opportunities, and incentives to transition to alternate employment by fishers and fisher households who diversify their job portfolio according to social and ecological forces of change. Fishing regulations, ENSO events, market integration, wealth and income of fisher households, and the expanding pulse of tourism in the Galapagos are related to the formation of alternate household livelihood alternatives of fishers to include a mix of tourism and fishing as circumstances and opportunities dictate.

From a livelihood perspective, we find that labor mobility through fishing in the Galapagos is enabled by social connections, wealth, education, and retraining programs that strengthened weak human capital. Former fishers have higher economic security and more adaptive capabilities than active fishers. Active fishers with lower formal education are experiencing a constriction of adaptive pathways out of fishing. Recent shifts through fishing show a willingness to leave without higher wages elsewhere. Adapting into a smaller fishing footprint means accepting small, "purposeful" adjustments and more socially, and politically "transformative" adjustments.

Further, this research highlights how adaptive capacity is shaped by the synergistic interactions between governance and local communities. Fishers understand that they cannot continue past practices unaltered (Ospina 2006), and they are independently pursuing a myriad of options to compensate for the growing difficulty in making a living. However, management decisions have placed ceilings on certain tourism livelihoods, such as the number of people allowed to become tourism operators to regulate tourism (El Universo 2009; Epler 2007). While the concern over tourism's unknown environmental capacity is necessary, a way forward must also be found for the Galapagos National Park (GNP) to weigh its regulatory responsibilities more synergistically, as well as more democratically with local communities.

Several issues described above have emerged through a synthesis of our work broadly conducted to understand human-environment interactions, explored within a coupled human-natural systems context, relying upon agent-based models. In our work in northeast Thailand, Ecuadorian Amazon, and the Galapagos Islands, we examine scenarios of change from baseline conditions, relying upon emergence, critical thresholds and triggers, and non-linear dynamics to assess shifts in system behavior. We also examine important social-ecological processes, for instance, population migration and tourism, household wealth and assets, job diversification, and alternate household livelihood strategies as well as land cover/land use change patterns and the social-ecological drivers of change. While the characterization of patterns has been useful and important to us, we are fundamentally interested in enhancing our process understanding. Configuring the model so that change can be readily visualized and understood can be a challenge, particularly, given scale dependencies, space-time lags, and exogenous dynamics. Feedback mechanisms and their dynamics are critically important as human behavior shapes the environment and the environment reshapes human actions. Varying seed points, deviating from the mean field perspective to examine "outlier" conditions, and searching for model convergence from a pallet of outcome possibilities are used to generate robust results and to enhance modeling interpretation and meaning. Our work continues, particularly, as we explore the behavior of fishers in the Galapagos Islands of Ecuador that are impacted by the pushes and pulls of alternate employment options, demographic characteristics, employment goals and intentions, and the vagaries of the Galapagos Marine Reserve – management policies (e.g., fishing quotas), disturbance regimes (e.g., ENSO events), and market conditions. Living with uncertainty in our models may limit their application in management and decision-making, although scenario testing offers important contributions to understanding "what if" alternatives to complex and non-linear dynamics.

References

Ben-Yami M (2001) Managing artisanal fisheries of Galápagos. A consultancy report–07-01-2001–03-02-2001. Charles Darwin Research Station, Santa Cruz, Galápagos

Chambers R, Conway G (1992) Sustainable rural livelihood: practical concepts for the twenty-first century. IDS discussion paper no. 296. University of Sussex, Institute of Development Studies, Brighton

Cinner JE, Daw T, McClanahan TR (2009) Socioeconomic factors that affect artisanal fishers' readiness to exit a declining fishery. Conserv Biol 23(1):124–130

El Universo (2009) Denuncia de influencias en los cupos turísticos para Galápagos. El Universo [Guayaquil] 19 July 2009, Noticias, País. http://www.eluniverso.com/2009/07/19/1/1447/DA25C09651BB42A087C97EFA9D154737. Last accessed 11 Apr 2011

Ellis F (1998) Household strategies and rural livelihood diversification. J Dev Stud 35(1):1–38

Ellis F (2000) Rural livelihoods and diversity in developing countries. Oxford University Press, Oxford/New York

Engie K (2015) Adaptation and shifting livelihoods of small-scale fishers in the Galapagos Marine Reserve, Ecuador. PhD dissertation, Department of Geography, University of North Carolina at Chapel Hill, Chapel Hill

Engie K, Quiroga D (2014) The emergence of recreational fishing in the Galapagos Marine Reserve: adaptation and complexities. In: Denkinger J, Vinueza L (eds) The Galapagos marine reserve: a dynamic social-ecological system; Galapagos Good Series; Social and ecological interactions in the Galapagos Islands (Walsh SJ, Mena CF (eds)). pp 203–226. Springer, New York

Epler B (2007) Tourism, the economy, population growth, and conservation in Galapagos. Charles Darwin Foundation, Puerto Ayora, Santa Cruz Island

Gonzalez JA, Montes C, Rodríguez J, Tapia W (2008) Rethinking the Galapagos Islands as a complex social-ecological system: implications for conservation and management. Ecol Soc 13(2). http://dlc.dlib.indiana.edu/dlc/handle/10535/2456

Grenier C (2007) Conservation contra natura: las islas Galapagos. Abya Yala, Quito, Ecuador

Haglund E, Ndjeunga J, Snook L, Pasternak D (2011) Dry land tree management for improved household livelihoods: farmer managed natural regeneration in Niger. J Environ Manag 92(7):1696–1705

Hearn A (2008) The rocky path to sustainable fisheries management and conservation in the Galápagos Marine Reserve. Ocean Coast Manag 51(8–9):567–574

Hearn A, Martinez P, Toral-Granda VM, Murillo JC, Polovina J (2005) Population dynamics of the exploited sea cucumber Isostichopus fuscus in the western Galápagos Islands, Ecuador. Fish Oceanogr 14(5):377–385

Henderson S, Dawson TP (2009) Alien invasions from space observations: detecting feral goat impacts on Isla Isabela. Galápagos Islands with the AVHRR. Int J Remote Sens 30:423–433
Honey M (2008) Ecotourism and sustainable development: who owns paradise? Island Press, Washington, DC
Instituto Nacional de Estadistica y Censos (INEC) (2010) Encuesta de condiciones de vida, Galápagos 2009–10. Instituto Nacional de Estadistica y Censos, Quito, Ecuador
Kerr SA (2005) What is small island sustainable development about? Ocean Coast Manag 48(7–8):503–524
Kerr S, Cardenas S, Hendy J (2004) Migration and the environment in the Galápagos: an analysis of economic and policy incentives driving migration, potential impacts from migration control, and potential policies to reduce migration pressure. Motu Economic and Public Policy Research, Wellington
Mahon R, McConney P, Roy RN (2008) Governing fisheries as complex adaptive systems. Mar Policy 32(1):104–112
Malanson GP, Walsh SJ (2015) ABM: individuals interacting in space (invited commentary). J Appl Geogr 56:95–98
Manson SM (2001) Simplifying complexity: a review of complexity theory. Geoforum 32(3):405–414
McCleary AL (2013) Characterizing contemporary land use/cover change on Isabela Island, Galápagos. In: Walsh SJ, Mena CF (eds) Science and conservation in the Galapagos Islands. Springer Science + Business Media, New York, pp 155–172
Mena CF, Walsh SJ, Frizzelle BG, Malanson GP (2011) Land use change of household farms in the Ecuadorian Amazon: design and implementation of an agent based model. J Appl Geogr 31(1):210–222
Michener WK, Baerwald TJ, Firth P, Palmer MA, Rosenberger J, Sandlin EA, Zimmerman H (2003) Defining and unraveling biocomplexity. Bioscience 51(12):1018–1023
Miller BW, Breckheimer I, McCleary AL, Guzman-Ramirez L, Caplow SC, Walsh SJ (2010) Using stylized agent-based models for population-environment research: a case from the Galapagos Islands. Popul Environ 31(6):401–426
Miller ML, Carter RW, Walsh SJ, Peake S (2014) A conceptual model for studying global change, tourism and the sustainability of iconic national parks. The George Wright Forum, A special issue on “Global Change and the World’s Iconic Protected Areas,” (Eagles P, Taylor PA) 31 (3):256–269
Nash S (2009) Ecotourism and other invasions. Bioscience 59(2):106–110
Nuñez RB (2007) Galápagos: La difícl relación entre lo demográfico y lo ambiental. In: Ospina P, Falconí C (eds) Galápagos: Migraciones, economía, cultura, conflictos, y acuerdos. Corporación Editora Nacional, and Universidad Andina Simón Bolívar, Quito, Ecuador, pp 59–74
Ospina P (2006) Galapagos, naturaleza y sociedad: actores sociales y conflictos ambientales en las Isla Galapagos, vol 55. Corporacion Editora Nacional. Quito, Ecuador
Percy MS, Schmitt SR, Riveros-Iregui DA, Mirus BB (2016) The Galapagos archipelago: a natural laboratory to examine sharp hydroclimatic, geologic, and anthropogenic gradients. WIREs Water. https://doi.org/10.1002/wat2.1145
Pizzitutti F, Mena CF, Walsh SJ (2014) Modeling tourism in the Galapagos Islands: an agent-based model approach. J Artif Soc Soc Simul 17(1):1–25
Pizzitutti F, Walsh SJ, Rindfuss RR, Reck G, Quiroga D, Tippett R, Mena CF (2016) Scenario planning for tourist management: a participatory and system dynamics model applied to the Galapagos Islands of Ecuador. J Sustain Tour. https://doi.org/10.1080/09669582.2016.1257011
Pollnac RB, Pomeroy RS, Harkes IHT (2001) Fishery policy and job satisfaction in three Southeast Asian fisheries. Ocean Coast Manag 44(7–8):531–544
Qin H (2010) Rural-to-urban labor migration, household livelihoods, and the rural environment in Chongqing Municipality, Southwest China. Hum Ecol 38(5):675–690

Quiroga D (2009) Crafting nature: the Galapagos and the making and unmaking of a "natural laboratory". J Polit Ecol 16(1):123–140
Rozzi R, Massardo F, Cruz F, Grenier C, Muñoz A, Mueller E (2010) Galapagos and Cape Horn: ecotourism or greenwashing in two emblematic Latin American archipelagoes? Environ Philos 7(2):1–32
Schuhbauer A, Koch V (2013) Assessment of recreational fishery in the Galapagos Marine Reserve: failures and opportunities. Fish Res 144:103–110
Scoones I (1998) Sustainable rural livelihoods: a framework for analysis. 72. Institute for Development Studies, Brighton
Stacey L, Fuks V (2007) Struggling for the golden egg: conservation politics in the Galapagos. Master's thesis. Available from Roskilde University Digital Archive
Taylor JE, Hardner J, Stewart M (2008) Ecotourism and economic growth in the Galapagos: an island economy-wide analysis. Environ Dev Econ 1:139–162
Terkla DG, Doeringer PB, Moss PI (1988) Widespread labor stickiness in the new England off-shore fishing industry: implications for adjustment and regulation. Land Econ 64(1):73–82
The International Ecotourism Society (TIES) (2016) What is ecotourism? Retrieved from https://www.ecotourism.org/what-is-ecotourism
The Nature Conservancy (TNC) (2017) What is ecotourism? Eco-tips and travel. Retrieved from https://www.nature.org/greenliving/what-is-ecotourism.xml
UCOOPEPGAL (2012) Galápagos Fishermen denounce thee violation of our constitutional rights by authorities of the Galápagos National Park. Pto Baquerizo Moreno
United Nations Educational, Scientific and Cultural Organization (UNESCO) (2010) Reactive monitoring mission report. Galapagos Islands 27 April–6 May 2010. Paris
Villacis B, Carrillo D (2013) The socioeconomic paradox of Galapagos. In: Walsh SJ, Mena CF (eds) Science and conservation in the Galapagos Islands: frameworks & perspectives; Social and ecological interactions in the Galapagos Islands. Springer, New York pp 69–85
Walsh SJ, Mena CF (eds) (2013a) Science and conservation in the Galapagos Islands – frameworks & perspectives. Springer Science & Business Media, New York, 243p
Walsh SJ, Mena CF (2013b) Perspectives for the study of the Galapagos Islands: complex systems and human-environment interactions. In: Walsh SJ, Mena CF (eds) Science and conservation in the Galapagos Islands: frameworks & perspectives. Springer, New York pp 49–67
Walsh SJ, Mena CF (2016) Interactions of social, terrestrial, and marine sub-systems in the Galapagos Islands, Ecuador. Sackler Colloquium on Coupled Human and Environmental Systems (Social Sciences, Environmental Sciences, and Sustainability Science). Proc Natl Acad Sci 113(51):14536–14543
Watkins G, Cruz F (2007) Galapagos at risk: a socioeconomic analysis of the situation in the archipelago. Charles Darwin Foundation, Puerto Ayora, Galápagos
Wilen JE, Stewart M, Layton DF (2000) Economic analysis of the Galapagos Marine Reserve resource management plan. Final report, University of California, Davis
Winterhalder B, Lu F, Tucker B (1999) Risk sensitive adaptive tactics: models and evidence from subsistence studies in biology and anthropology. J Archaeol Res 7(4):301–347
World Wildlife Fund (2003) Migration and environment in the Galapagos Islands, Quito, Ecuador

Chapter 6
Socioecological Systems and the Management of the Natural Resources in the Galapagos

Diego Quiroga

Introduction

The more than 30,000 people living in the Galapagos Islands have had to adapt to a rapidly changing physical and social global reality. As is the case of other socioecological systems that depend on the use of natural resources, the sustainability of the system is limited by the erosion of the fragile resources. In the Galapagos, different social sectors have arrived to use and, in some cases, exploit these natural resources with negative effect on the unique ecosystems and species. As in other oceanic islands, these social groups have created emergent solutions to their most pressing needs; in some cases these solutions are sustainable in the long-term, but many have also shown to be detrimental for the survival of the societies that created them (Whittaker 1998).

The idea of the sustainability of emergent solutions is challenged by the concept of the tragedy of the commons developed by Garrett Hardin in 1968. Regulations and regulators have tried, with mixed results to shape the relations between the local residents and their environment and to monitor and prevent some of the threats to the native and endemic species. Recently the work of some social scientists has explored the conditions under which the tragedy can be adverted without a reliance only on top-down solutions. We need to understand under which conditions systems that emerge to use resources for fishing, tourism, and other activities are sustainable in the long term or to what extent some degree of government intervention is necessary.

In this chapter we explore the multiple levels at which decisions are made and the effectiveness of these embedded processes. We explore the way in which the play

D. Quiroga (✉)
Universidad San Francisco de Quito USFQ, Colegio de Ciencias Sociales y Humanidades, Quito, Ecuador
e-mail: dquiroga@usfq.edu.ec

T. Kvan, J. Karakiewicz (eds.), *Urban Galapagos*, Social and Ecological Interactions in the Galapagos Islands, https://doi.org/10.1007/978-3-319-99534-2_6

between emergence bottom-up systems and government control (top-down) generate a dynamic and often conflictive relationship between groups and individuals with diverse interests and objectives.

Most of the permanent residents who live in the Galapagos depend directly or indirectly in the use of the unique and fragile natural resources. Tourism, fisheries, conservation, and science place alternative and often conflictive claims to the fragile territories that make the Galapagos ecosystems. Through the Galapagos National Park (GNP), the national government regulates the use of the common pool resources (CPR), but often these regulations are questioned and challenged by the residents. Alternative, illegal, and often unsustainable use of the resources creates tensions between the different users and the GNP. Ultimately the resilience of the system must be guaranteed with effective and sustainable use of the resources through novel management systems.

Human Settlement

The Galapagos presents a unique set of geophysical and biological conditions that have allowed the fauna and flora of the area to evolve. There are several geographical and biological characteristics that make this archipelago unique in the world. One is the fact that these oceanic islands are at an ideal distance from the mainland (1,000 Km) which allows some animals and plants to arrive but not others; furthermore, the islands are sufficiently isolated from the mainland for evolutionary processes to create new species. A similar situation occurs with the distance between islands as animals and plants become isolated from each other for further processes of evolution to take place. The existence of a series of oceanic currents provide nutrients and create different environments in the islands which explains why many of these islands have animals and plants assemblages that are usually not found in equatorial latitudes, such as penguins, corals, tropical fish, and sea lions (Valle 2013). The contrasting diversity and biological processes that are found in these islands have made the Galapagos a world famous place for the study and the analysis of the evolution of the fauna and the flora.

In contrast to other oceanic archipelagos of the Pacific, the Galapagos was never subject to an early process of human colonization. There is little of any evidence of people colonizing or even visiting the islands before the Spaniards arrived in the twenty-first century. This probably helps to explain why a large proportion of the native animals and plants can still be found in the islands. For several centuries, Europeans and North Americans, including British privateers and American whalers, visited the islands and extracted resources such as sperm whales, fur seals, and tortoises (Gonzalez et al. 2008). These visitors contributed to the extinction or the near extinction of some of these species (Latorre 2001).

In 1832, shortly after the foundation of Ecuador as a country, a group of colonists were sent by the Ecuadorian government to mark a presence of Ecuador in the archipelago. This population of 200 to 300 people led by José María Villamil established the first colony in Floreana (Latorre 2001; Gonzalez et al. 2009). Another group of Ecuadorians also attempted to establish themselves in that same island and failed. It was only some years later, in the 1880s, that the first successful colony was established in San Cristobal. At first the settlers were mostly interested in agriculture and they engaged in the production of goods for their own consumption and for international trade. Coffee, sugar cane and derivatives, cattle, and other products became the most important exports of the islands (Latorre 2001).

By the beginning of the twentieth century, the total population of the Hacienda El Progreso was of some 400 people. Juan Manuel Cobos, the enterprising Ecuadorian who brought the people and the machinery to establish a large sugar mill in San Cristobal, saw the Galapagos as an ideal place to generate legal and illegal commerce and to establish industry in the pursuit of development and progress. His vision and action made the Galapagos a frontier for Ecuadorian colonization and development. He established a center of commerce and industrial development far from the oversight of mainland Ecuador where he could rule over the land and people. Other colonies were also started on the islands, such as on Isabella by Ecuadorians and by Europeans in several of the other islands. The establishment of these colonies was a reason for rapid transformation of the Galapagos in the agricultural zones on few of the higher islands where rain and soils allow for agriculture to develop.

Charles Darwin arrived to the islands in September of 1835, and although his short visit to the Galapagos was not the unique event that transformed his vision immediately, the collections, observations, and the conversations he had in the archipelago became important points in the creation of his theory of evolution (Quiroga and Sevilla 2016). The various species of mocking birds, tortoises, and finches and their distribution in the islands became the basis for the revolutionary and controversial theory of evolution. The idea that order in the form of complex forms and social systems emerges from the struggle for survival of different individual organisms was the basis for the mechanisms that resulted in the idea of evolution by natural selection. These ideas were inspired to some extent by the work of the Scottish liberal tradition of Adam Smith and Bernard Mandeville and other conservative authors such as Malthus (Quiroga 2016) who also underlie many of the theories in emergent and complex systems (Levin 1998). The Galapagos became an ideal biological laboratory for the study of evolution in the last part of the nineteenth century and the beginning of the twentieth century as scientist like Louis Agassiz and the California Academy of Science arrived to look for evidence to prove or disprove the Darwinian conception of biological evolution (Larson 2001).

Regulation and Emergence

In the second half of the twentieth century, besides the Galapagos being established as a center for science to understand the evolution of species, it was also a center for tourism and conservation. Fisheries and tourism gained importance during after the

1970s and this caused many people to move to the coastal areas and the port towns where these new activities were based. It was during these decades that the port towns played an important role in the new configuration of the islands as they became a hub for tourism, fisheries, and commerce. Towns grew at a surprisingly fast rate between the 1990s and the early 2000s some years reaching growth rates of 6% or higher (Kerr et al. 2004). By 2010, the census indicated that there were more than 25,000 residents in the Islands (Villacis and Carrillo 2013). A new law for the Galapagos was passed in 1998 that established among other things regulations for migration to the island. The creation of this new law placed a restriction on the number of people coming to the islands. The number of residents in the Islands is as of this writing probably more than 35,000 people living permanently in the archipelago. More than 50% of the economic activity in the islands is now based on tourism (Watkins and Cruz 2007). At the beginning, most of the tourism was based on cruise ships that travel between the islands; however increasingly tourism is now based on land operations based on the towns that demand more goods, energy and services (Epler 2007). As a result, more residents and newcomers are involved in the service economy mostly around tourism and fewer in fisheries and agriculture, and there has been a shift not only in the economy but also in the vision and culture of many of the people living in the Archipelago (Quiroga 2013).

The growth in the number of people coming to the island and their raising standard of living has meant that Galapagos is one of the provinces of Ecuador with the greatest consumption per capita. This increase in consumption and the decrease in local production of some of the basic products have meant that there is a greater demand for goods coming from the outside. Connectivity with the mainland has increased, and ships and airplanes are arriving more frequently to the islands; currently there are at least one ship a week and around ten flights arrive daily bringing not only people, including locals, national visitors, and international tourists, but also products from the mainland (Pizzitutti et al. 2016). The increase in ships and airplanes arriving has meant that there is also a greater risk for the introduction of foreign species, for oil spills, and new diseases that can affect the people, the animals, and the plants. There is an ever-growing demand for fossil fuels not only because of the needs of the cruise boats but also as the tourist who are now increasingly staying in towns demand air conditioning, refrigeration for food products, and transportation on land and sea. Although alternative energy generation projects have been implemented, in particular wind and solar, these projects are insufficient to satisfy the growing demand. In the twenty-first century, as jobs in the service industry including tourism, conservation, and public sectors increased, there has been an abandonment of the agricultural and fishing activities. This abandonment has meant that there is greater reliance on imported food products.

The national government manages the Galapagos through different institutions which include the Galapagos Government Council and its president, who has the status of a minister, is appointed by the national government, and reports directly to the Ecuador's President and the director of the Galapagos National Park who is under the minister of natural resources. There are also representations of other ministries such as tourism, education, agriculture, health, electricity, and renewable energy and other sectors of the government that play a role in different aspects of the Galapagos. The roles and functions of these different institutions often overlap. For example, although the Galapagos National Park is in charge of regulating tourism and the activities of the large and small boats that operate in the Galapagos Marine Reserve (GMR), the tourism ministry is also involved in monitoring and regulating some of these activities. Besides the presence of the national government, there are other different local governments (Gobiernos Autonomos Descentralizados) led by elected authorities. These include three municipalities and several Juntas Parroquiales (governing rural areas). There are three mayors for each of the main towns: Santa Cruz, San Cristobal who is also in charge of the small town in Floreana, and Isabela.

The Galapagos National Park regulates and manages most of the land area (almost 97%) and all the marine areas (the Galapagos Marine Reserve). Many of the emergent human activities such as fishing, tourism, agriculture, gathering, and hunting infringe of the terrestrial or the marine areas of the park. During the last few decades, these emergent activities have included new tourism activities such as sport fishing, construction of hotels and residential houses, and the introduction of new transportation systems such as speed boats for fishing and tourism. As these novel activities developed and their impact is understood, the government, through the GNP, attempts to implement regulations to mitigate their negative effects. Thus, the activities of the GNP impact upon and reduce the economic activities of the residents of the Galapagos. There have been several conflicts between conservation and growth, in particular when it comes to development on the coastal areas including the building of municipal infrastructure such as roads and different tourism activities that the GNP does not approve.

Fisheries and tourism have thus generated bottom-up emergent processes, some of which have resulted in the tragedy of commons such as the overfishing of lobster and sea cucumber and the uncontrolled tourism in some places (Hearn 2008). Emergent processes have also produced conservation initiatives such is the formation of the Galapagos Marine Reserve, which was the result of the efforts of by different sectors including fishermen, conservationists, and the tourism sector to eliminate the entrance of the industrial fishing fleet from the mainland of Ecuador and other countries.

The Dynamics Play Out

In his 1968 article Garrett Hadin talks about the deterioration of open access areas due to the overexplotation (Hardin 1968) Elionor Ostrom and her team have mentioned eight different ways in which communities can sustainably manage their commons in an emergent and self-organized process (Ostrom 1990). The creation and management of the marine reserve in 1990, which was at the time of its creation one of the largest reserves in the world, was to some extent an interesting way of managing the commons. Its creation and management involved the participatory management board composed of different sectors such as tourism, science, conservation, and fisheries that played a role in the allocation of the marine resources. The marine reserve was created at a particular moment when people were united in the effort to prevent the overexploitation of the resources of the marine areas. Though far from perfect, the process resulted in people discussing their different issues and negotiating agreement on important aspects of the management of the GMR.

The creation of the marine and terrestrial protected areas required the establishment of laws and regulations. These regulations in many cases have not been sufficient to prevent the tragedy of the commons. In the Galapagos, there are many examples of the failure of directed and top-down process to prevent the threat to the endemic organisms. These include the introduction of species and the overexploitation of resources such as overfishing of sea cucumbers and lobsters, illegal harvesting of shark fins, the overuse of resources for tourism, and uncontrolled expansion of the urban areas.

In general, one can say that there have been some bottom-up processes that have resulted in self-organized management solution, but many of these processes have been hampered by the government. In some cases the bottom-up process has led to revisions of the regulations after they were enacted. Such is the case with those local people who have a permit to take tourists to places close to the towns. Many of the permits for day tours were given to people who had been fishermen. To get a license to become a day tour operator, the fishermen had to resign their fishing license. In the case of San Cristobal, there were a few permits given to take tourists to iconic places like Kicker Rock which is a snorkeling and diving site close to town (40 min boat ride). After the regulations were passed and the permits were issued, pressure from the day tour operators made it possible for them to have access to Española, an island that is 2 h south of San Cristobal and which is now part of the day tours from San Cristobal. Many local people monitor each other's operations to prevent overuse and destruction of the new visiting areas. Several regulations in Galapagos as in other areas are the result of bottom-up and top-down processes.

Another example is what happened with the fishers who were interested in taking tourists fishing to complement their income. It was originally suggested by some fishers that they could take tourists as a way of diversifying their activities and decrease the fishing effort. This activity originally called *pesca vivencial* was an activity in which tourists would participate and share a day with a fisherman (Schuhbauer and Koch 2013, Engie and Quiroga 2013). The municipality of San Cristobal promoted at the same time sport fishing, an activity that was not authorized by the GNP. Finally both ideas have been partially accepted and a hybrid product emerged with few boats approved to be part of what was called pesca artesanal vivencial (PAV). Some fishers have now specialized in "catch and release" fishing of billfish, while others, or in some cases the same boats, also go fishing with tourist for smaller fish closer to the shores. Limits of 50 pounds per catch have been established, and many fishermen have developed new types of itineraries where the catches are kept at a minimum. In the case of San Cristobal, lobbying by local fishermen who are involved in PAV has meant that new "resting places" were open so the fishers can take the tourist to visit these places. These new resting places has meant that the fishing activity is being transformed into a new tourism operation, and fishermen now offer tours such as the 360, a circumnavigation of the island on which typically no fishing takes place. On other islands like Santa Cruz and Isabella, PAV has now become a form of tourism, which in many ways is a new adaptive and emergent process with some top-down regulation.

Whereas the creation of the GMR included significant participation of the local community and many of the developments described above have been instigated by the local people, there are also examples of many top-down regulations with little local participation. A clear example of this top-down process was the creation in 2016 of the marine sanctuary. The sanctuary was the result of the decision and plans coming from the Charles Darwin Station, National Geographic, and the Government of Ecuador without much knowledge on the part of most of the fishermen or other members of the local population. The sanctuary was implemented supposedly to protect the sharks on the northern part of the archipelago. This is ironic, however, as the sharks were already protected in the GMR for decades, which makes many people suspicious that the creation of the sanctuary was only a media stunt to improve the image of Rafael Correa, the national president at that time, whose environmental record was being questioned by many because of his policies in the Amazon forest where he promoted oil drilling and mining. Others maintain that the main purpose of the sanctuary was to promote tourism in an area managed by a few companies that are based outside of the Galapagos and that have little local participation on their operations. Fishermen, many of whom say that they were never consulted, are upset by the new sanctuary which impinges on their fishing areas for lobster and other pelagic and demersal fisheries. The approval of the use of long lines as a pilot study, which in the past was prohibited because of the relatively high levels of bycatch, is seen by some as part of the negotiation with the fishermen to lower their dissatisfaction with the creation of the marine sanctuary.

As we have shown, the regulation of the Galapagos terrestrial and marine protected areas is the result of both top-down and bottom-up processes. Some emergent process such as the fishing of sea cucumbers in the 1990s and early 2000s resulted in the tragedy of the commons for those resources, whereas others show that local people can manage their common pool resources in a sustainable manner. Imposed top-down regulations have resulted in some cases in the overuse of resources and serious environmental and social impacts. In some cases, this dynamic tension between emergent and top-down processes has been successful in organizing and distributing the spaces and resources of the Galapagos. Such is the case of the tourism operations around the island that are part of the land-based island hopping tours. In its beginning, the management of the GMR was an interesting although not fully successful effort to create participatory systems of management. Ironically, more recent experience has been top-down with little consultation and participation of the local stakeholders, which has led to tensions and conflicts.

Conclusion

In this chapter I have described the way in which emergent and top-down process have been transforming the Galapagos and shaping its environment. The management of the islands and its unique and precious resources has been the result of the dynamic and unpredictable processes that result from the tensions generated by the interaction between the emergent and at times unsustainable actions of the local people and the equally dynamic and often unsuccessful top-down regulations imposed by the management authorities. Despite many errors and some failures, the Galapagos can be seen as a complex adaptive system that is learning from past mistakes and generating new sustainable solutions. It is an open question whether the different actors, including the local people, external interest groups, and the authorities regulating the system, will be able to act in time to shape the direction this complex system is taking and prevent a collapse of many ecosystems that are unique to these islands and are the bases for a flourishing tourism industry.

Understanding the emergent processes through complex adaptive system theory can give us some clues about the points at which intervention is required. When a system is reaching a tipping point as described by the theory, it can transition to a more complex and adaptive system, it might remain as it is, or it might pass through a destruction phase. Most regulatory work in the islands has focused on incremental maintenance of the status quo, but the review above of the past 300 years illuminates that the island system has been tipped into instability by human intervention and the status quo is unstable. The challenge here is what transformations might be needed to a better outcome and how a governance system might oversee such a transformation.

References

Engie K, Quiroga D (2013) The emergence of recreational fishing in the Galapagos Marine Reserve: adaptation and complexities. In: Denkinger J, Vinueza L (eds) The Galapagos marine reserve (Social and ecological sustainability in the Galapagos). Springer, New York, pp 203–226

Epler B (2007) Tourism, the economy, population growth, and conservation in Galápagos, vol 68. Charles Darwin Foundation, Santa Cruz, Galápagos

González J, Montes C, Rodríguez J, Tapia W (2008) Rethinking the Galapagos Islands as a complex social-ecological system: implications for conservation and management. Ecol Soc 13:13

Hardin G (1968) The tragedy of the commons. Science 162:1243–1248

Hearn A (2008) The rocky path to sustainable fisheries management and conservation in the Galápagos Marine Reserve. Ocean Coast Manag 51(8–9):567–574. https://doi.org/10.1016/j.ocecoaman.2008.06.009

Kerr S, Cardenas S, Hendy J. (2004) Migration and the environment in the Galapagos: an analysis of economic and policy incentives driving migration, potential impacts from migration control, and potential policies to reduce migration pressure. Motu working paper 03–17. Motu Economic and Public Policy Research, Wellington

Larson E (2001) Evolution's workshop: god, science on the Galapagos. Penguin Books, London

Latorre O (2001) La Maldición de la Tortuga: Historias Trágicas de las Islas Galapagos (Cuarta Edi). Artes Grá fi cas Senal,Impresenal Cia, Quito, Ecuador, p 226
Levin SA (1998) Ecosystems and the biosphere as complex adaptive systems. Ecosystems 1(5):431–436
Ostrom E (1990) Governing the Commons: The evolution of institutions for collective action. Cambridge University Press, New York
Pizzitutti F, Walsh SJ, Rindfuss RR, Gunter R, Quiroga D, Tippett R, Mena CF (2016) Scenario planning for tourism management: a participatory and system dynamics model applied to the Galapagos Islands of Ecuador. J Sustain Tour 25(8):1117–1137. https://doi.org/10.1080/09669582.2016.1257011
Quiroga D (2013) Changing views of the Galápagos. In: Walsh SJ, Mena CF (eds) Science and conservation in the Galapagos Islands, Springer, New York, pp 23–48
Quiroga D (2016) Darwinian emergence, conservation. In: Quiroga D, Sevilla A (eds) Darwin, Darwinism and conservation, 2016. Springer, New York, pp 135–150
Quiroga D, Sevilla A (2016) Darwin's Galapagos myth; Darwin, emergent process, and the conservation of Galapagos ecosystems. In: Quiroga D, Sevilla A (eds) Darwin, Darwinism and conservation, 2016. Springer, New York, pp 1–7
Schuhbauer A, Koch V (2013) Assessment of recreational fishery in the Galápagos Marine Reserve: failures and opportunities. Fish Res 144:103–110
Valle CA (2013) Science and Conservation in the Galapagos Islands. In: Walsh SJ, Mena CF (eds) Science and conservation in the Galapagos Islands, Springer, New York, pp 1–22
Villacis B, Carrillo D (2013) The socioeconomic paradox of Galapagos. In: Walsh SJ, Mena CF (eds) Science and conservation in the Galapagos Islands. Springer, New York, pp 23–48
Watkins G, Cruz F (2007) Galápagos at risk: a socioeconomic analysis of the situation in the archipelago. Charles Darwin Foundation, Puerto Ayora, Galápagos
Whittaker RJ (1998) The human impact on islands ecosystems: the lighthouse keeper's cat and other stories. In: Island biogeography: ecology, evolution and conservation. Oxford University Press, Oxford, pp 228–256

Part III
Models of Change

Chapter 7
A Model-Based Approach to Study the Tourism Sustainability in an Island Environment: The Case of Galapagos Islands

Paola A. Espin, Carlos F. Mena, and Francesco Pizzitutti

Introduction

One of the biggest challenges for humanity today is to deal with the consequences of natural environment drastic changes generated around the planet by anthropic activities. People interact at different time and space scales with the Earth's natural environment producing multi-scale changes (Hollings and Sanderson 1996) that in turn affect the mere possibility to maintain and sustain the human societies' well-being. In coupled humans and natural systems (CHANS), two intrinsically complex systems, the natural ecosystems and the human societies, interact reciprocally, forming a supersystem that shows the typical marks of the complexity: non-linearity, feedback loops, and emergent and hierarchical organization, among others (Liu et al. 2007). As usual in complex systems, the study of CHANS cannot be limited to the description of isolated parts in a pure mechanistic way. A holistic, multidisciplinary approach is required to treat and represent all the CHANS ecological and social aspects and to take into account the intricate network of feedbacks and interactions.

In this paper we present a system dynamics (SD) (Forrester 1961) approach to the study of a particular type of CHANS: an insular touristic destination like the Galapagos Islands where in the last decades the explosive tourism development posed serious treats to the integrity of the irreplaceable insular ecosystems and also

P. A. Espin (✉) · C. F. Mena
School of Biological and Environmental Sciences, Universidad San Francisco de Quito, Ecuador and Galapagos Science Center, San Cristobal Island, Galapagos, Ecuador
e-mail: cmena@usfq.edu.ec

F. Pizzitutti
School of Biological and Environmental Sciences, Universidad San Francisco de Quito, Ecuador and Galapagos Science Center, San Cristobal Island, Galapagos, Ecuador

Duke University, Durham, NC, USA

T. Kvan, J. Karakiewicz (eds.), *Urban Galapagos*, Social and Ecological Interactions in the Galapagos Islands, https://doi.org/10.1007/978-3-319-99534-2_7

to the well-being of local Galapagos populations (Ministerio de Relaciones Exteriores 2007; UN 2007). The limited geographical and demographic extent and the rich interactions occurring at different time and space scales make the Galapagos archipelago a perfect natural laboratory where to study the complex interaction between natural and human environments (Quiroga 2009).

Islands represent 5% of Earth's surface (Towle 1985) with over 100,000 islands supporting 20% of global biodiversity (IUCN 2015). The combined characteristics of isolation and scale make many islands ecologically and culturally unique. On the other hand, these same characteristics represent a set of challenges to Islands populations in terms of limited resource availability, weak resource security, and limited natural carrying capacity (Deschenes and Chertow 2004). Furthermore, islands fragile environment is not only subject to internal pressures but also to external forces coming from the larger system in which it exists. Island ecosystems and oceanic archipelagos are highly vulnerable to human interventions that could transfer biological invasions that will diminish endemic species or to global economic forces that prioritize short-term exploitation of resources over traditional resource use practices (Kerr 2005; McElroy and de Albuquerque 1990; Parque Nacional Galapagos 2009a, b).

The rich and diverse ecosystems of many islands have made them very attractive to develop tourism activities despite the irreversibility human's impacts could have on islands native habitats. An integrated understanding of the complex implications that nature tourism has on island's sustainability deserves a deeper attention and studies. Islands around the world have the higher portion of recorded species extinctions (80% of known extinctions (IUCN 2015)) making it a major concern to adequately manage the pressures on these fragile and vulnerable ecosystems.

The Galapagos Islands are one of the most well-known examples of islands with extremely rare endemic species where tourism is blooming as an economic driver threatening its vulnerable ecosystems (González et al. 2008). Pressures associated with tourism, introduction of invasive species, demographic growth, and illegal fishing and governance issues threaten to destroy the endemism of the Islands and its unique biodiversity (Mena et al. 2013). Nature tourism in Galapagos is the main driver of economic growth. It employs directly, or through indirect employment in the support sectors, 40% of the resident population and represents 65.4% of the local economy (Kerr et al. 2004). Between 1999 and 2005, the GDP of the Islands increased 78% from a base of 41 million of US$, giving to the Islands an annual GDP growth rate of 10%. Tourism explains 74% of this growth, contributing not only to the Islands economy but also to the national economy via the archipelago demand for goods and services from Ecuador's mainland (Brewington et al. 2012). Despite this growth, average income per capita in the Islands has increased by only 1.8% annually, and residents complain of receiving few benefits from tourism. Tourism has been a main driver of population growth in the archipelago through immigration of work force from mainland, which made the Ecuadorian government to include restrictions on immigration in the Galapagos Special Regime Law in 1998 (González et al. 2008). Currently, the slow growth in GDP per capita is creating political pressures to further explore development options either through commercial fishing in the nature reserve or through increasing the annual numbers

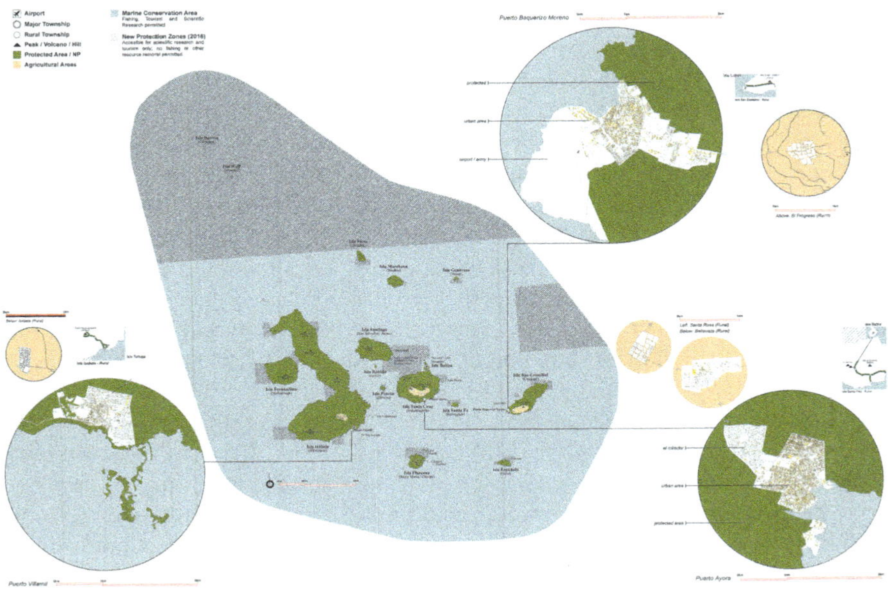

Fig. 7.1 The Galapagos Islands

of tourists (Taylor et al. 2006); both threaten to further exacerbate the pressures on the Islands' ecosystems (Fig. 7.1).

For more than three decades, management efforts in Galapagos were focused on regulating activities in the protected area, with little attention to what was happening with the human settlements on the colonized land. Nowadays, there is an increasing awareness of the issues arising by rapid population growth (Lu et al. 2013), which requires a holistic approach to conservation and management of the archipelago as an entire unit and not on a component basis: national park, marine reserve, and inhabited land. Socioeconomic and biological factors need to be recognized and incorporated in order to have a comprehensive management plan for the archipelago (Epler 2007).

Within this context, the aim of the study is to provide a systemic assessment of tourism dynamics and its implications for the sustainability of sensitive island ecosystems to better inform academia and policy-makers that are responsible of defining long-term strategies to promote tourism without harming social or environmental capitals.

System Dynamics

Tourism is an extremely complex dynamic phenomenon (Xing and Dangerfield 2011), which cuts across many sectors and encompasses a set of social, economic, and environmental relations. Tourism complexity needs to be understood in an integrated manner in order to better define strategies that can assure the long-term sustainability of local communities and the quality of the environment, particularly in small islands and ecologically sensitive touristic destinations (Gössling 1999; Hunter and Green 1995; UNWTO 2015).

The SD method looks to answer questions about the dynamic nature of complex systems by analyzing the causal structure that generates behavioral patterns over time. It focuses on the internal structure to explain problematic behaviors rather than focusing on external disturbances or random events. System dynamics captures complexity by recognizing nonlinear relations and accounting for lagged relations among variables. To be able to explain the factors defining a system's structure, they incorporate concepts from any discipline with emphasis in biophysical and social interactions. It is important to highlight that unlike econometric models, the main focus of analysis in system dynamics models is the general tendencies over time, whether the system is stable, unstable, growing, declining, oscillating, or in equilibrium, rather than precise numeric results (Meadows 1980; Sterman 2000). As Xing and Dangerfield (2011) point out, system dynamics allows visualizing tourism as a network of integrated systems, whose performance can be inferred through quantification and computer simulation. Explicit mapping and analysis of feedback stimulates an intuitive grasp of dynamics and the debate among audience to improve system's design.

There has been an increase in the number of publications using system dynamics to study tourism phenomena in the last decade (Sedarati 2015). System dynamics models are purpose oriented (Sterman 2000), and due to tourism system's complexity, the spectrum of areas covered by the publications is broad among tourism sectors: transportation, environment, recreation, travel trade, cultural activities, and accommodation (Sedarati 2015). Most of the papers apply system dynamics to address specific problems; this is the most suitable application of the method as it facilitates setting the boundaries for discussion and fulfilling data requirements (Sterman 2000). Patterson et al. (2004) choose to use system dynamics to model the environmental, social, and economic systems of Dominica, which makes the model broad and very data demanding. This is not a method limitation but rather author's choice. In this research, authors had access to all the required data; this can be a limitation for other case studies where no local data is monitored or when resources for primary data collection are limited. The broad approach of the model also makes tourism to loose relevance in the analysis, by making it just another component of the economic system.

Several papers are leveraged on generic structures to explain models' relationships, for example, Georgantzas (2003) use of Sterman's supply chain model and Hernández and León's (2007) use of the predator-prey structure. This can be considered a model's strength, as these are already well-known proven elements of structure that adequately capture the dynamics they look to explain. This could also facilitate understanding and models' acceptance when readers are familiarized with them. Furthermore, it contributes to building a generic system dynamics toolbox set that can be used across domains (Lane and Smart 1996).

In terms of tourism sustainability assessment, several studies emphasize some dimensions of sustainability to the extent that it is relevant. In this sense, they do provide valuable sustainability insights that could be useful in other locations: Semeniuk et al. (2010) approach to explore interactions between tourists and stingrays, for example, can be applied in other ecotourism destinations where interactions between tourists and animals are part of the location's attraction. Patterson et al. (2004) approach of modeling the system is the one that provides the closest look at an integrated assessment of sustainability's social, environmental, and economic dimensions. However, as already mentioned, the broad perspective makes tourism loose relevance in the analysis, and no specific indicators are developed to assess tourism sustainability.

Tourism and Population and Labor in Galapagos

Population

The first population census in 1962 reported 2,391 inhabitants in the Galapagos Islands. By 2010, this value has multiplied by more than ten times to 25,124 inhabitants (INEC 2013). The average annual growth rate of the population in Galapagos over the last two decades has been 4.83%, while the same rate for total Ecuador was 2.03% (INEC 2014a). The rapid and sustained population growth has been primarily attributed to the inflow of tourism dollars that attracted Ecuadorian immigrants looking for employment opportunities. National economic crisis during the 1980s and 1990s, including the financial collapse that led to the dollarization of Ecuadorian economy in 1999, also motivated Ecuadorians from mainland to migrate to Galapagos, one of the only local spots where labor was demanded (Epler 2007). On the other hand, national policies have also contributed to increase immigration pressure. People living in Galapagos benefit from a set of subsidies including energy, airfares, and shipping which has increased the standards of living in the islands above those in many parts of Ecuadorian mainland (Kerr et al. 2004). According to Kerr et al. (2004), half to two thirds of population growth in Galapagos is related to immigration (around 3.5–4%), partly offset by emigration (around 1–2%).

Tourism

As noted above, over the years, the number of tourists visiting Galapagos has grown rapidly to a current level of 224,755 tourists arriving in 2015, a 226% increase compared to 2000 figures (68,856 tourists (Observatorio de Turismo Galapagos 2015)). This trend has exceeded any expected behavior from authorities, conservationist, or local population. In the 1980s official guidelines recommended to restrict the number of tourists to 12,000 per year. Later on, a government commission evaluated the situation and raised the limit to 25,000. Today, the parameter of groups at the same time is applied by the Galapagos National Park Service to establish an accepted load of visitors per day in the natural visiting sites. Based on its approximation, 1,702 visits per day are allowed in sites that can only be visited on cruise mode and 1,960 visits per day in sites that can be reached by land-based tourists (Mena et al. 2013). This accounts for approximately 248,677 visits per year. The fraction of the number of the load of tourists visiting Galapagos in the same day (tourists at the same time) over the number of local inhabitants also shows an increase over time (5.7% in 1990, 6.1% in 2001, and 10.1% in 2010). This value reflects that the number of tourist arriving to Galapagos is growing at a faster pace than the growth of local residents.

Over the years, the tourism industry has continued to grow and diversify, and currently a broad range of products and services are offered to tourists wanting to visit the islands. Visitors can opt between two main forms of tourism: the boat-based and the land-based tourism. Cruise-based tourism is advocated as less invasive, as it requires fewer infrastructures to be developed on land (Epler 2007). The data on land accommodation has been collected by the Galapagos authorities since 2007; the trend for the last 9 years shows a decrease in onboard tourism of −1.6% on average per year, both in foreign and domestic tourists. This means that the increase of tourism experienced in Galapagos in this period has been mainly driven by the land-based mode. Land tourism represented 91% in 2007, and in 2015 it was 97% of total domestic tourism; for foreign tourism this proportion has changed from 30% in 2007 to 54% of total foreign tourism in 2015. These changes reflect the increasing importance that land tourism is having on the islands.

Labor Market in Galapagos

Tourism growth has been a main driver of immigration to Galapagos. The size of the labor market in Galapagos measured by the economically active population[1] as a fraction of the working age population[2] was 63% by 2010, while in Ecuador's mainland this proportion was 50.8% in the same year (INEC 2014b) meaning that among the people in the working age, a bigger portion is either working or willing to work in Galapagos.

The employment[3] rate in Galapagos has also been constantly higher in comparison to Ecuador. Over the three censual periods -1990, 2001, 2010- (INEC 1990,

[1] According to Ecuador National Institute of Statistics and Census (INEC), economically active population is defined as people aged 10 and older who are employed or those who are unemployed but are available for working and seeking employment (INEC 2015).

[2] INEC defines working age population as all people aged 10 years or older (INEC 2015).

[3] According to INEC, employed are people aged 10 and older who are engaged in any activity to produce goods or provide services in exchange of remuneration or benefits (INEC 2015).

2001, 2010a), Galapagos reflects an employed population of around 90%, while in Ecuador this rate remains below 46%; this explains why Ecuadorians find much easier to get a job in Galapagos and tend to migrate from Ecuador mainland. As Kerr et al. (2004) identified, immigration is likely to occur from provinces that have low employment to provinces that have higher employment rates, as people are more likely to get employed or find better wage opportunities in higher employment areas. Furthermore, salaries have been reported to be higher in Galapagos. According to the survey of living conditions made in 2009 (INEC 2010b), the average monthly income for an employee in Galapagos was US$ 772, while in Ecuador mainland it was US$ 252. This value is partly offset by the cost of living that is higher in Galapagos. According to the survey, the cost of a basic basket of goods and services in Galapagos was US$ 863, while in Ecuador mainland it was US$ 378.

From an analysis of the structure of the labor market in Galapagos, it is possible to see that in the Islands, tourism[4] activities have a higher relative importance (19%) than in Ecuador mainland (11%), while agriculture, cattle, and fishing are more important in Ecuador (22%) than in Galapagos (10%). Important to notice is the relevance that the public sector has as an employer in the islands, representing 11% of the employed population, while in Ecuador mainland it represents only the 4%. The process of decentralization of power to the province, the strengthening of the municipalities, and the creation of the Galapagos National Institute (INGALA) have increased public employment in the islands (Kerr et al. 2004). The market structure of Galapagos along the three periods of censual data shows an increase of the tourism employment sector, while agriculture and fishing show a decrease. As Kerr (2005) highlights, island economies can quickly become highly specialized, to the virtual exclusion of all unrelated activities. Between 2001 and 2010, tourism employment grows by 27%, while agriculture and fishing decrease by 24%. This process is due to the fact that tourism has an impact on the local farmer preferences, giving to them an opportunity to improve their income, while agriculture is a low profit sector. Farmers leave the rural areas and move to the urban centers at the port-side. The abandonment of agricultural land increases the risk of expansion of invasive species and an additional threat to Galapagos food security as less food is produced locally (Mena et al. 2013). For fishing, the official reports might be underestimated since it only considers people that report fishing as its main economic activity and illegal fishing is not reported although widely diffused. In this line, Kerr et al. (2004) report that many local people that work in tourism-related activities are also involved into fishing – at least part time – especially sea cucumber during the season since it requires little skills and the profits are high.

[4]Tourism is not measured directly as an economic activity in the National Census. This value aggregates the data reported as lodging, restaurants and food services, transport, and communications.

The System Dynamics Model

On the basis of the information presented in the previous section, Table 7.1 summarizes the main trends and issues for the relevant variables that are represented in the model. This section will describe the conceptual structure, or dynamic hypothesis, that explains how these variables are related and what the consequences of these relationships are.

The dynamic hypothesis and the relationship among variables of our model are represented in a causal loop diagram (CLD), which also serves as a guide for the quantified modeling effort. Special attention is placed on the emergence of feedback loops, either reinforcing[5] or balancing,[6] which are relevant in determining the behavior of the system in relation to tourism sustainability in Galapagos over time. All systems consist of networks of reinforcing and balancing feedbacks, and all dynamics arise from the interaction of these loops with one another (Sterman 2000).

As showed in the CLD presented in Fig. 7.2, the growing patterns in the reference modes of population and tourism can be explained by several reinforcing feedback loops. The first set of loops are the three *tourism-population growth loops* that describe the dynamics of tourism, labor demand, migration, and population in Galapagos. An increase in the number of tourists arriving to Galapagos translates in a higher demand of labor in tourism and related sectors. Higher demand of labor, from one side, reduces labor availability[7] which increases the job attractiveness of the island to immigrants looking for employment (loop 1). On the other hand, it reduces unemployment rate, which has two migration effects. It increases immigration (loop 2) from mainland Ecuadorians looking for jobs, and it reduces emigration (loop 3) of local residents that have fewer incentives to move out of the Islands. The effect of unemployment in migration is relative to the unemployment rate in Ecuador's mainland; if Ecuador has lower unemployment than Galapagos, mainland Ecuadorians will have fewer incentives to immigrate and vice versa; this effect is represented by the unemployment ratio. The impact of the three loops in migration rates increases population. More population reduces the revenues per capita that local residents perceive from the expenditures that tourists make in the Islands. As tourism revenues per capita decrease, the social pressure to expand tourism increases as a way to increase tourism revenues again with the arrival of more tourists (*tourism expense balancing loop*).

The second set of reinforcing loops are the two *labor demand-immigration loops* and the *labor demand-emigration loop*. This set of loops explains the effect that local population has on labor demand and consequently on the migration rates, further influencing population. More residents in Galapagos increase the demand for

---

[5] Positive or reinforcing feedback loops tend to reinforce or amplify whatever is happening in the system. It denotes a self-reinforcing process (Sterman 2000).

[6] Negative or balancing feedback loops counteract and oppose change. It denotes a self-correcting process (Sterman 2000).

[7] Equals to the supply/demand ratio of the labor market. For the modeling exercise, it is labor force/desired labor.

Table 7.1 Main trends and issues for the relevant variables represented in the model

Variable	Main trends	Issues
Population	Continuous increase in population, annual growth rate of 4.8% on average over the last two decades (1990–2010) More than 60% of people living in Galapagos were born outside the Islands People that moved to Galapagos in the last 5 years have increased 2% on average per year between 1990 and 2010	Population growth driven by immigration from Ecuadorian mainland attracted by employment in tourism and related activities Additional pressures fostering immigration: Ecuador's financial crisis and national policies of subsidies for Galapagos residents Population growth increases the pressures on islands resources (water and food production, waste and sewage management) and the risk of introduction of invasive species
Tourism	Increase of tourists visiting Galapagos, annual average growth rate of 8% on the last 15 years The fraction of tourists at the same time over local population increases from 5.7% in 1990 to 10% in 2010 Decrease in cruise-based tourism (−1.6% per year), increase in land-based tourism (8.5% per year) on the last 9 years Average stay on cruise-based tourism has increased 2.4% between 2007 and 2012 and 5.5% on land-based tourism	Tourists' numbers almost reaching the current established limit of visits in the natural sites Tourism generates pressures in terms of introduction of invasive species, demographic growth, illegal fishing, and governance issues which threaten to destroy Island ecosystems Cruise-based tourism is advocated as less invasive, as it requires fewer infrastructures to be developed on land Longer stays are associated with immigration growth, increase in inter-island traffic, and the introduction of more invasive species
Labor market	The relative size of the labor market is higher in Galapagos than in Ecuador's mainland (50.8% vs. 63.2%) The employment rate in Galapagos has been constantly higher than in Ecuador (90% vs. 46%) Tourism and public sector are important employers in Galapagos Tourism shows an increasing trend as an employment sector (+27% between 2001 and 2010); agriculture and fishing decrease as an employment sector (−24% between 2001 and 2010)	Galapagos has a very dynamic labor market with high employment rate and an increasing demand of labor which attracts migrants from Ecuador mainland Tourism has changed the structure of the labor market; less people are employed in agriculture and fishing since tourism-related activities are seen as more profitable Abandonment of agriculture increases the risk of expansion of invasive species and the dependency of the Islands on food imports

labor across all sectors as more goods and services are demanded and a bigger public sector is needed to manage population affairs. As already explained, demand for labor stimulates immigration through a reduction of labor availability (loop 1) and lower relative unemployment (loop 2). Lower unemployment also reduces emigration rate. The changes in migration rates further increase local population. The *labor supply-immigration balancing loop* counteracts immigration increase given

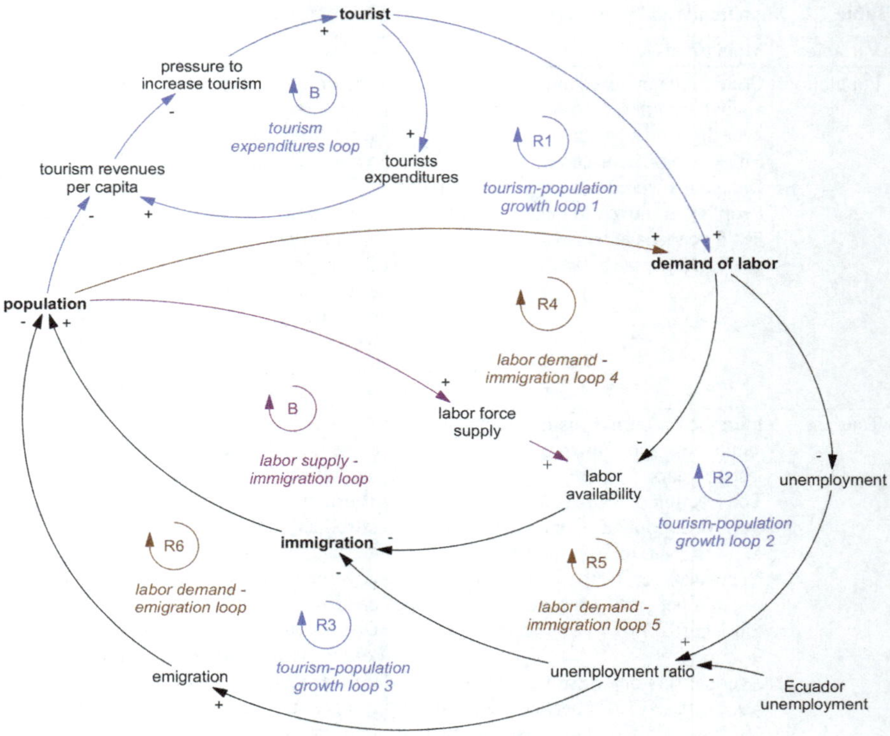

Fig. 7.2 CLD with tourism, population, and labor dynamics

that more population also translates in higher labor supply which increases labor availability and reduces the job attractiveness of the Islands.

From this analysis it is important to point out the two main forces impacting migration to and from Galapagos. From one side, the *internal dynamics* of Galapagos' labor market are represented by the labor availability ratio, which reflects the balance between labor supply and demand and its subsequent impact on immigration. On the other side, the *external dynamics* of the labor market are represented in the unemployment ratio, which affects immigration and emigration rates based on a comparison of the labor market in Galapagos relative to Ecuador mainland.

In spite of the growing trends created by this set of reinforcing feedback loops, it is not possible to have unlimited growth in a finite system, moreover in island ecosystems, where issues of isolation and scale translate into a sensitive environment with limited resource availability, weak resource security, and limited natural carrying capacity. The natural limitations of the ecosystem come to play to counteract the reinforcing loops through several balancing loops that are connected to the natural attractiveness erosion of the Islands. Over time, an increase in the erosion of natural attractiveness due to tourism and population growth will reduce the number of tour-

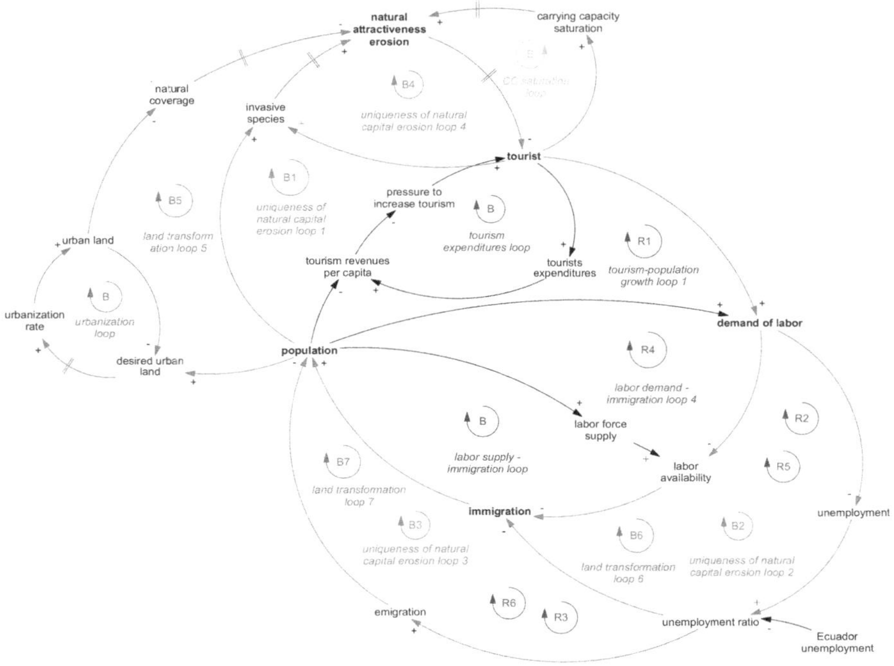

Fig. 7.3 CLD with natural limit balancing loop

ist arrivals to Galapagos given that its natural habitats are the main attraction of the archipelago. Figure 7.3 adds these loops to the ones previously explained.

The first set of balancing loops are the four *uniqueness of natural capital erosion loops* that explain the increasing risk that more population, driven by migration dynamics, and more number of tourists have in the introduction and spreading of invasive species. More invasive species over time will erode natural attractiveness through the extinction of endemic species and habitat destruction. The higher the number of tourist arrivals, the higher the risk of undermining the quality of natural capital, which, in the long term, will reduce the attractiveness of the location and in consequence the number of tourists. Over time, less tourism will reduce the immigration rates as the attractiveness of Galapagos as an employment destination will also decline; however, the impacts associated with the introduction of invasive species cannot be reversed, which further highlights the need to adequately manage the long-term tourism strategy in the Galapagos Islands.

The dynamics of urbanization and land use changes are captured in a second set of balancing loops: the three *land transformation balancing loops* and the *urbanization loop*. The *urbanization loop* reflects the impact that population increase has in the urbanization rate. The more people migrate to the urban areas in Galapagos, the higher the need to extend urbanization; when more land is urbanized, the shortage of urban plots and the desired urban land are also reduced. On the other hand, the three *land transformation loops*, two passing through immigration and one through emigration, capture the negative impact that higher urbanization has on the quality of natural capital by reducing the natural land coverage. As already explained, over time natural capital erosion will reduce the attractiveness of the location for tourism, causing a reduction in the immigration rate and in the need for urbanization. However, urbanized land cannot be transformed back to natural ecosystems, which makes natural capital erosion from urbanization also irreversible.

Finally, the *carrying capacity saturation balancing loop* represents the dynamics related to the capacity of the Galapagos National Park to receive visitors. The more tourists arrive, the higher the saturation of the established carrying capacity will be. Over time, receiving more tourists than the actual capacity of the visiting sites in the Galapagos National Park will increase the natural attractiveness erosion, which will negatively impact the arrival of tourists.

The time horizon over which the reinforcing loops of tourism and population growth and the natural balancing loops operate is different considering that environment-related dynamics usually take longer periods than socioeconomic ones. It is for this reason that growth is currently the governing pattern of the system; however, in the long term, a shift in loop dominance could cause the system to collapse if adequate measures are not taken. The dynamics of these loops are represented in a quantitative model which facilitates the simulation of long-term scenarios to better understand the interactions of the main loops governing the system and assess the potential risk of a collapse and the implications of tourism for Islands' sustainability in the long term.

Discussion

This research relied on the use of formal SD model to approach the representation of the social and ecological implications of tourism in Island ecosystems. The model design was based on the methodological development of tool to understand and assess the dynamics of tourism sustainability in island ecosystems. The initial stage of the model development approach allowed identifying the problematic dynamics arising from tourism, which were subject of analysis in the later stages. This process included the selection and development of model-based indicators to assess tourism sustainability that were implemented as part of the model's structure to facilitate the creation of a dashboard view to visualize assessment results.

The application of the designed method to the selected case study allowed demonstrating the usefulness of integrating assessment indicators into a system dynamics model to understand the dynamics of tourism and assess its social and ecological implications. Likewise, this also allowed addressing the other questions about the structure and feedback loops that explain the impact of tourism growth in the demographic dynamics of island's population, the structure and main feedback loops that are responsible for the social and ecological pressures associated with tourism and demographic growth in island ecosystems, and the structure and key indicators that support the sustainability assessment of tourism strategies in island ecosystems to gain a balance between local population wealth and ecosystem protection.

As exposed, one of the main implications of tourism in relation to island's sustainability has to do with its impacts on population growth. This relationship was explained through a reinforcing feedback loop in which tourist arrivals increase labor demand fueling immigration of job seekers. Higher immigration increases population that over time requires more and more tourist arrivals to sustain social wealth. This reinforcing cycle of tourists and population growth intensifies natural capital degradation through the interaction with environmental balancing loops, which over time will affect the potential of the location to attract the demanded number of tourists to maintain social wealth. Tourism growth in Galapagos and consequent population increase accelerate the erosion of natural attractiveness. Island ecosystems, as the one in the Galapagos archipelago with rich and diverse habitats that developed in extreme state of isolation, are very sensitive to any kind of external pressure. Therefore the presence of human populations in these originally uninhabited areas represents a risk to disturb the native environment. The structure of the system was used to define a set of indicators that could be useful to assess the sustainability of tourism strategies in island's ecosystems. The analyzed dimensions included population, labor, environmental, and social dynamics. Each dimension was characterized by a set of indicators corresponding to the main variables of interest.

The internal dynamics of the labor market in Galapagos deserves a further, more detailed, analysis. For the Galapagos case, higher tourism has disruptive implications fostering employment across tourism-related sectors while reducing the activities in fishing and agriculture. Tourism-related sectors attract people from the rural areas given that tourism provides them with better income opportunities, in particular taking into account that local production has to compete with cheaper mainland import products. An increasing contraction of the fishing and agriculture sector threatens local food security and could also raise the risk of invasive species by increasing the dependence of the Islands on food imports (Pizzitutti et al. 2016).

The environmental implications of tourism were represented by the index of natural attractiveness erosion that aggregates introduction of invasive species, saturation of carrying capacity of the protected area, and urbanization and land use in the non-protected area. Any kind of human activity represents a risk to the endemism of Islands' ecosystems. The higher the tourism growth, the higher the level of introduced species since more tourism implies increasing mobilization of passengers and food imports to satisfy the demands of local residents and tourists. The higher the growth in tourism, the faster the Islands will reach a saturation in the installed carrying capacity of the visiting sites on the protected area. Finally, a higher growth in tourism arrivals and population also translates in a higher rate of urbanization and land transformation in the area outside the national park. Possibly the maximum target of occupied land will be reached, and further expansion of the populated area will be necessary, to the detriment of the GNP extension. The aggregated indicator of these three forces reflects the increasing natural attractiveness erosion as a consequence of tourism and population growth that over time feeds back and impacts the number of tourist arrivals.

The social implications of tourism in island's ecosystems have to be assessed through a social wealth index that aggregates the employment rate and tourism revenues per capita. This index is also affected and affects the tourism-population growth cycle. Although tourism revenues per capita increase once tourism arrivals grow, this increase decreases over time, and could even be reversed, due to the increase in population that results from higher tourism. When the economic benefits from the increase in tourism arrivals are diluted on a per capita basis, a feedback effect is created because local residents pressure to increase even more tourism arrivals looking to compensate for the loss of tourism revenues per capita.

Implications for Policy and Future Research

A validation process of the model by policy-makers and Galapagos stakeholders is needed to fully assess the usefulness of the SD approach to support planning and decision-making about tourism in island ecosystems. However the model presented here constitutes a significant attempt to understand that a solely measure will not improve tourism sustainability in island's ecosystems given the strong reinforcing and balancing feedbacks that are put in place by tourism. The relevance of tourism activity was highlighted since the major constraints to improve social and environmental goals were associated with tourism and population growth. Galapagos, as several other islands' destinations, relies on its natural ecosystems to attract visitors; thus, its protection needs to be prioritized for tourism to be sustainable in the long term. Tourism strategies should therefore aim to limit the influence of the tourism and demographic growth loop to avoid an overshoot and collapse process in which over time tourism will erode itself while destroying irreplaceable natural sites in the process.

Other complementary measures are also necessarily to balance social and environmental goals. A strategy to move from high-volume, low-value tourism to a different kind of tourism that maximizes its economic value with a lower environmental impact could compensate for social wealth loss. Furthermore, an improvement of technical and administrative capacity would be a prerequisite to guarantee an effective enforcement of any proposed actions.

The model-based approach that was developed is meant to provide insights into the mechanisms through which tourism influences demographic dynamics and exerts social and ecological pressures on island ecosystems. Thus it supports the critical arguments that when ecotourism operates under the mainstream paradigm of continuous growth, the expected social and environmental goals are not likely to be achieved. It also contributes to assess the sustainability implications of tourism strategies through the integration of a set of model-based indicators and aggregated indexes. This approach tackles the limitations of previous Galapagos representation failing to capture in an integrated way the socioeconomic and environmental dimensions to assess tourism sustainability in Island ecosystems. Nevertheless, future research can further develop such insights and assess their transferability and relevance to other contexts beyond the Galapagos case study.

Limitations of the model itself are also worth of further exploration. Model's boundaries could be extended to assess the impacts of changing national and international contexts in local tourism dynamics of island ecosystems. Implications of wealth distribution among residents should also be addressed since the benefits from tourism are usually not equally distributed among local population. Also, the spreading pattern of introduced species is far more complex than the approximations that could be put in place from a SD point of view. This aspect could be further explored by using other modeling techniques such as agent-based modeling to simulate emerging patterns from agent's interactions that could later be aggregated in the system dynamics model. From the discussions with the experts of the GSC team and the review of secondary sources, it is clear that the dynamics of fishing in Galapagos are very complex and could deserve to be a topic of analysis by its own. In addition, the official data reported does not include any statistics on illegal fishing of sharks and sea cucumbers. Previous conflicting episodes of strikes and threats on protected species reflect only some of the difficulties to regulate this sector.

Lastly, the development of model-based indicators to assess tourism sustainability could be refined through the evaluation of the aggregated indexes of social wealth and natural attractiveness erosion for different contexts. Additional variables could be included, for example, to represent social well-being instead of social wealth. Furthermore, the formulation and the aggregation method could include participatory approaches that allow capturing the perspective of relevant stakeholders regarding the importance of the variables aggregated and the degree of substitutability between them, looking to provide a more compelling and stronger tool.

These further developments will provide an opportunity for building over the contribution that was presented in this paper with regard to the advantage of applying a system dynamics model-based approach that allows understanding how tourism impacts the socioeconomic and ecological dynamics of island ecosystems from an analysis of the underlying feedback structure interacting. In addition, the application of the presented model-based approach is expected to support further development of a set of indicators to assess the sustainability of tourism strategies facilitating scenario analysis and policy testing.

References

Brewington L, Rosero O, Bigue M, Cervantes K (2012) The quarantine chain. Wildaid, San Francisco

Deschenes P, Chertow M (2004) An island approach to industrial ecology: towards sustainability in the island context. J Environ Plan Manag 47(2):201–217

Epler B (2007) Tourism, the economy, population growth, and conservation in Galapagos. Charles Darwin Foundation, Puerto Ayora

Forrester JW (1961) Industrial dynamics. The MIT Press, Cambridge, MA

Georgantzas NC (2003) Tourism dynamics: Cyprus' hotel value chain and profitability. Syst Dyn Rev 19:175–212

González JA, Montes C, Rodríguez J, Tapia W (2008) Rethinking the Galapagos Islands as a complex social-ecological system: implications for conservation and management. Ecol Soc 13(2)

Gössling S (1999) Ecotourism: a means to safeguard biodiversity and ecosystem functions? Ecol Econ 29:303–320

Hernández J, León C (2007) The interactions between natural and physical capitals in the tourist lifecycle model. Ecol Econ 62:184–193

Hollings CS, Sanderson S (1996) Dynamics of (dis)harmony in ecological and social systems. In: Floke K-GS, Mäler C (eds) Right to nature: ecological, economics, cultural, and political principles of institutions for the environment. Island Press, Washington, DC

Hunter C, Green H (1995) Tourism and the environment. A sustainable relationship? Routledge, London

INEC Instituto Nacional de Estadísticas y Censos (1990) V Censo de Población y IV de Vivienda. Retrieved from http://redatam.inec.gob.ec/cgibin/RpWebEngine.exe/PortalAction?&MODE=MAIN&BASE=CPV1990&MAIN=WebServerMain.inl

INEC Instituto Nacional de Estadísticas y Censos (2001) VI Censo de Población y V de Vivienda. Retrieved from http://redatam.inec.gob.ec/cgibin/RpWebEngine.exe/PortalAction?&MODE=MAIN&BASE=CPV2001&MAIN=WebServerMain.inl

INEC Instituto Nacional de Estadísticas y Censos (2010a) VII Censo de Población y VI de Vivienda. Retrieved from http://redatam.inec.gob.ec/cgibin/RpWebEngine.exe/PortalAction?&MODE=MAIN&BASE=CPV2010&MAIN=WebServerMain.inl

INEC Instituto Nacional de Estadísticas y Censos (2010b) Encuesta de condiciones de vida Galápagos 2009–2010. INEC, Quito

INEC Instituto Nacional de Estadísticas y Censos (2013) Principales características demográficas de Galápagos. Resultados del Censo 2010. Talleres gráficos del CGREG, Puerto Baquerizo Moreno, Galápagos

INEC Instituto Nacional de Estadísticas y Censos (2014a) Anuario de Estadísticas Vitales – Nacimientos y Defunciones 2014. INEC, Quito

INEC Instituto Nacional de Estadísticas y Censos (2014b) Encuesta nacional de empleo, desempleo y subempleo. Indicadores Laborales, INEC, Quito

INEC Instituto Nacional de Estadísticas y Censos (2015) Glosario de términos-ENENDU. INEC Instituto Nacional de Estadísticas y Censos: http://www.ecuadorencifras.gob.ec/glosario-determinos-enendu/. Accessed 8 Apr 2016

IUCN (2015, July 9) Island ecosystems. International Union for Conservation of Nature: http://www.iucn.org/about/union/commissions/cem/cem_work/tg_islands/. Accessed 19 May 2016

Kerr S (2005) What is small island sustainable development about? Ocean Coast Manag 48:503–524

Kerr S, Cardenas S, Hendy J (2004) Migration and the environment in the Galapagos: an analysis of economic and policy incentives driving migration, potential impacts from migration control, and potential policies to reduce migration pressure. Working paper series, 03–17

Lane DC, Smart C (1996) Reinterpreting 'generic structure': evolution, application and limitations of a concept. Syst Dyn Rev 12(2):87–120

Liu J, Dietz T, Carpenter SR, Alberti M, Folke C, Moran E et al (2007) Complexity of coupled human and natural systems. Science 317(5844):1513–1516. https://doi.org/10.1126/science.1144004

Lu F, Valdivia G, Wolford W (2013) Social dimensions of 'nature at risk' in the Galápagos Islands, Ecuador. Conserv Soc 11(1):83

McElroy JL, de Albuquerque K (1990) Managing small-island sustainability: towards a systems design. Nat Resour 26:23–31

Meadows D (1980) The unavoidable a priori. In: Randers J (ed) Elements of the system dynamics method. Pegasus Communications, Waltham, pp 23–57

Mena CF, Walsh, S., Pizzitutti F, Reck G, Rindfuss R, Orellana D, Toral V, Valle C, Quiroga D, García J C, Vasconez L, Guevara A, Sanchez M E, Frizelle B, Tippett R (2013) Determinación de las relaciones sociales, ambientales y económicas que permitan desarrollar, en base a procesos de modelación, potenciales escenarios de sostenibilidad del sistema socio-ecológico de las Islas Galápagos con énfasis en la dinámica del flujo de visitantes al archipiélago. Ministerio del Ambiente Ecuador, Parque Nacional Galápagos. Galapagos Science Center

Ministerio de Relaciones Exteriores (2007) UNESCO inscribe a las islas Galápagos en la lista de patrimonios en peligro. Ministerio de Relaciones Exteriores y Movilidad Humana: http://www.cancilleria.gob.ec/unesco-inscribe-a-las-islas-galapagos-en-la-lista-de-patrimonios-en-peligro//. Accessed 1 June 2016

Observatorio de Turismo Galapagos (2015). http://www.observatoriogalapagos.gob.ec/. Accessed 7 Apr 2016

Parque Nacional Galapagos (2009a) Conservación y restauración de ecosistemas insulares: Control y erradicación de animales introducidos. Parque Nacional Galápagos Ecuador: http://www.galapagospark.org/nophprg.php?page=parque_nacional_especies_invasoras_animales. Accessed 31 May 2016

Parque Nacional Galapagos (2009b) Conservación y desarrollo sustentable: Gestión ambiental en áreas pobladas. Parque Nacional Galápagos Ecuador: http://www.galapagospark.org/nophprg.php?page=desarrollo_sustentable_zonificacion. Accessed 2 June 2016

Patterson T, Gulden T, Cousins K, Kraev E (2004) Integrating environmental, social and economic systems: a dynamic model of tourism in Dominica. Ecol Model 175:121–136

Pizzitutti F, Walsh S, Rindfuss R, Gunter R, Quiroga D, Tippett R, Mena C (2016) Scenario planning for tourism management: a participatory and system dynamics model applied to the Galapagos Islands of Ecuador. J Sustain Tour 25(8):1117–1137

Quiroga D (2009) Crafting nature: the Galapagos and the making and unmaking of a "natural laboratory". J Political Ecology 16(1):123

Sedarati P (2015) System dynamics in tourism: a systematic literature review. Master thesis dissertation, Universidade do Algarve, Algarve

Semeniuk C, Haider W, Cooper A, Rothley K (2010) A linked model of animal ecology and human behavior for the management of wildlife tourism. Ecol Model 221:2699–2713

Sterman J (2000) Business dynamics. Systems thinking and modeling for a complex world. McGraw-Hill, Boston

Taylor JE, Hardner J, Stewart M (2006) Ecotourism and economic growth in the Galapagos: an island economy-wide analysis. University of California, Department of Agricultural and Resource Economics. Giannini Foundation of Agricultural Economics, Davis

Towle EL (1985) The island microcosm. In: Clark J (ed) Coastal resource management: development case studies. Research Planning Institute, Columbia

UN (2007) Galápagos Islands added to UNESCO list of endangered heritage sites. UN News Centre: http://www.un.org/apps/news/story.asp?NewsID=23036&Cr=heritage&Cr1=#.V1VuEJPhCRt. Accessed 1 June 2016

UNWTO (2015) Why tourism matters. World Tourism Organization UNWTO: http://www2.unwto.org/content/why-tourism. Accessed 16 May 2016

Xing Y, Dangerfield B (2011) Modelling the sustainability of mass tourism in island tourist economies. J Oper Res Soc 62:1742–1752

Chapter 8
Toward Urban Self-Sufficiency in the Galapagos Islands

Justyna Karakiewicz

Context

Previous chapters in this volume have outlined aspects of Galapagos social and economic development that identify some of the major challenges for the future of the islands as inhabited places. The growth of tourism holds promise as a means to drive further development yet also brings dangerous consequences by exposing the fragile environment to pollution, particularly from fossil fuels and lubricants but also from waste. Basic provisions of food and water must be addressed, but there is also a need for meaningful engagement in the tourist economy and the development of a *galapagueña* future. The earlier chapters have also described how the local population has addressed the mandated reduction of fishing by positively engaging in knowledge-based income-generating activities that work with the needs of conservation and environmental protection through both top-down and emergent responses. Finally, theoretical approaches have been presented in the earlier chapters with which specific urban futures can be postulated and paths for transition be explored. This chapter takes up these themes and experiences to explore changes to urban form that might support a different economy and greater self-reliance to support a resilient future.

Laws and policies have been enacted to guide human settlement and protect the fragile environment of the islands. More recently these have included the creation of the National Park in 1959, Galapagos Special Laws and the Regional Management Plans in 1998, and the Management Plan for Protected Areas of Galapagos for Good Living in 2013–2014. While these have had positive impact on the islands in the past 50 years, the problems have not disappeared. Unfortunately, despite conservation

J. Karakiewicz (✉)
Melbourne School of Design, Faculty of Architecture, Building and Planning, University of Melbourne, Parkville, VIC, Australia
e-mail: justynak@unimelb.edu.au

T. Kvan, J. Karakiewicz (eds.), *Urban Galapagos*, Social and Ecological Interactions in the Galapagos Islands, https://doi.org/10.1007/978-3-319-99534-2_8

policies, migration controls, and restrictions on land development, the degradation of the islands is accelerating, and so far no initiatives have been identified to reverse this process (Bensted-Smith 1999; Pizzitutti et al. 2014).

This history has been mapped onto Holling's cycle of adaptive change to show how major events through the history of the Galapagos Islands have led to regime shifts and allowed Galapagos reconfigured itself to a new system (Gonzales et al. 2008). This depiction suggests that the islands are today in a unique position where there is no evident structure that emerges, despite several attempts from the government, National Park, or national special laws, which were passed in 1998 (Fig. 8.1).

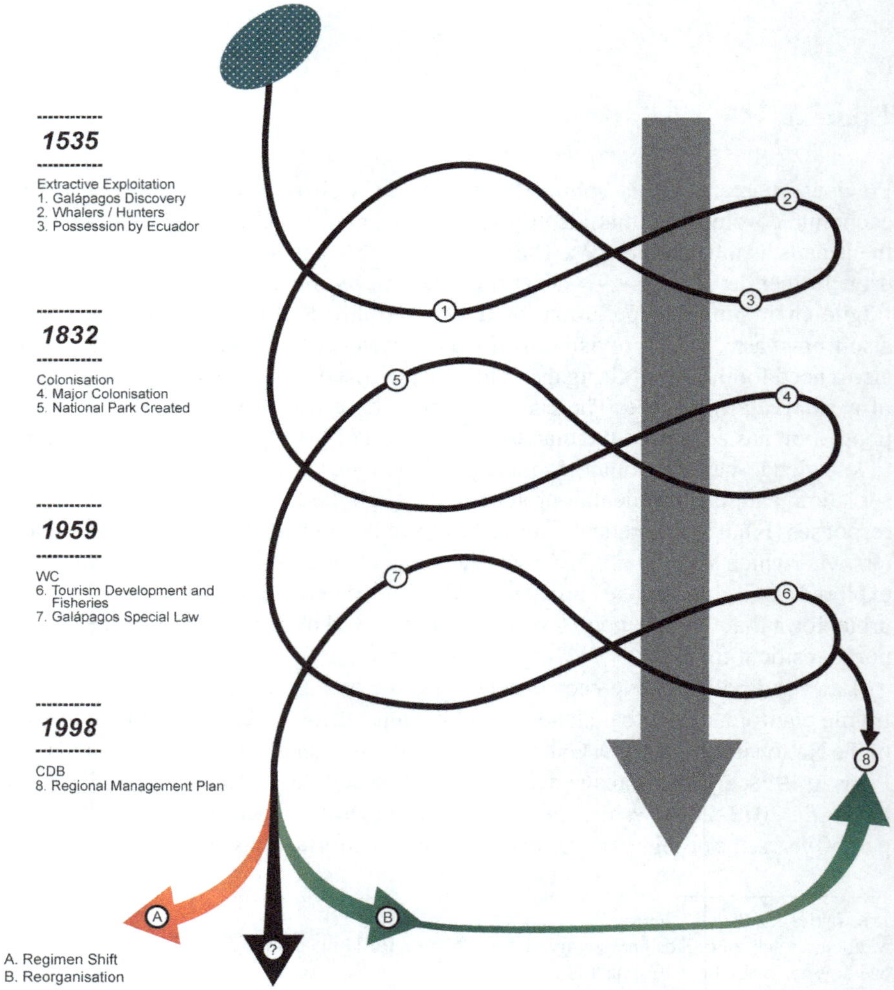

Fig. 8.1 Cycle of adaptive change on the Galapagos Islands. (Redrawn based on Gonzales et al. 2008)

Furthermore, there is clear evidence that the system may shift into undesirable stage.

Social change is problematic, and those who are least advantaged typically bear the burden of the change. Walsh et al. in Chap. 5 in this book document the effect of the rise of tourism on fishers. It needs to be noted that 11% of the population in the Galapagos live in extreme poverty (lacking access to two or more basic needs) and a further 40% are also classified in poverty in the Unsatisfied Basic Needs Index, that is, lacking one basic need (Granda León et al. 2013). The levels of poverty in rural areas are higher (78%) than in the urban areas (47%). However, poverty rates in the islands are lower than the average across Ecuador, and therefore the islands continue to be attractive to internal migration from the mainland. Most of the newcomers find employment in tourism, and, as Gonzales et al. (2008) note, tourism will continue to play a major role in the future of the Galapagos Islands.

> Our resilience-based approach to the Galapagos situation reveals that tourism is, and will probably continue to be, the most important driver of change through its direct effects on the archipelago's ecology, economy, and socio-cultural system. Energy flows, growing economic opportunities, and positive economic feedbacks associated with the current model of tourism have the potential to accelerate major changes and threaten the sustainability of the archipelago in the near future. In its first stages, the model of cruise-boat tourism appeared to be a good choice for Galapagos, but after the rapid generalization of locally based tourism and the increasing presence of infrastructure and land-based services such as hotels, restaurants, and transportation, the economic growth driven by tourism became somewhat of a double-edged sword. On the one hand, it contributes to environmental conservation and improvements in the standard of living through taxes and admission fees; on the other hand, it overheats the islands' economy, creating new economic niches and opportunities that drive immigration and unsustainable growth. If the present crisis is to be solved, the identification of the most appropriate tourism model for Galapagos should be a top priority of research and planning for the next few years. (Gonzales et al. 2008)

There is considerable work still to be undertaken to respond to the challenges identified in their paper. Here we will address some aspects of the opportunities in the urban context to engage residents more directly in the tourist experience while enhancing the quality of life.

The problems facing Galapagos are enormous, and hoping that we will be able to solve them is arrogant and pointless idea. However, we can try to develop a framework, which could lead not only to better understand of what is actually happening but also through this process try to build some knowledge that may lead to a paradigm shift or small change in people's perception and attitudes toward a more sustainable and more resilient future. For that we first of all need to ask four fundamental questions:

1. Do we have structure in place?
2. Do we have process in place?
3. Do we have resources?
4. Do we have set of values to make changes possible?

We undertook a thought exercise through design to examine these questions and identified Puerto Baquerizo Moreno as an appropriate laboratory to consider how small changes can lead to reconfiguration of the whole system toward a more resilient and sustainable future for the coupled natural and human systems. The research has been carried out in annual design studios on the island of San Cristobal from 2013 to 2016. This chapter reports on work conducted during the last in which I worked with Jose Holguin of the University of San Francisco de Quito and 11 students.[1] The studios allow us to test some hypotheses and present them to both the residents and the government to provoke discussion of alternative ways forward.

Case Study of San Cristobal, Puerto Baquerizo Moreno

The capital city of the Galapagos Islands is Puerto Baquerizo Moreno, located on San Cristobal, the easternmost island of the archipelago (Fig. 8.2). With population of 7,000 people, it is the second biggest settlement in Galapagos. Although the

Fig. 8.2 Puerto Baquerizo Moreno

[1] Eight students were from the University of Melbourne, Raynaldo Ali, Hilton Hei Tong Chau, Isaac Zhiren Chen, Sovina Chow, Tommy Aik Meng Heng, Shenia Lay, Cathy Zhenchen Tang, and Jeanie Tsz Lam Tang, and three from the University of San Francisco de Quito, Alejandro Ramos Alban, Stephania Villalba Andrade, and Claudia Robalino.

Fig. 8.3 Typical urban structure at Puerto Baquerizo Moreno. One of the few timber buildings in 2014; it was replaced in 2015

population of Puerto Baquerizo Moreno is small, the metro area is quite significant, the settlement covering 848 sq. km. Densities therefore are very low, with many of the urban plots not occupied, abandoned, or transformed into rubbish tips. While initial construction was of timber and lava rock, most of the timber-structured buildings have been replaced by concrete or brick building, built with material imported from the continent (Fig. 8.3).

Aligning our work with that of the Galapagos Science Center, our work has used CAS theory to examine urban change (Karakiewicz et al. 2015; Karakiewicz 2016). Holling (2001) articulated the concepts of the cycle of change and of panarchy to replace the formalism of "hierarchies" with less deterministic top-down modes of development. The cycle is useful to describe a process of change for one ecosystem function operating over a time period. The adaptive cycles are characterized by three properties: wealth (the inherent potential of a system that is available for change), controllability (the degree of connectedness between internal controlling variables and processes), and adaptive capacity (the degree of the resilience of the system). Looking at Fig. 8.4, we can see clearly that at stage α, the system is characterized by the greatest uncertainty but also the greatest resilience. Restructuring and innovation is only visible in this phase, which is different to the common believe that stability and certainty should be a bases for resilience. In Fig. 8.5 we have interpreted Holling's characteristics of adaptive cycles as a panarchy model for our work.

The panarchy model allows for the description of interconnected cycles operating in varying time periods, nesting shorter cycles into longer cycles. This panarchy model allows us "to rationalize the interplay between change and persistence, between predictable and unpredictable" (Gunderson and Holling 2009). It combines

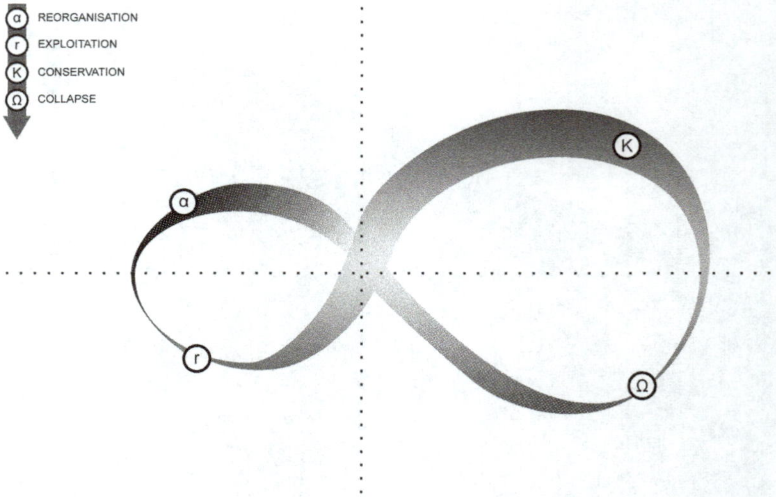

Fig. 8.4 Cycle of adaptive change

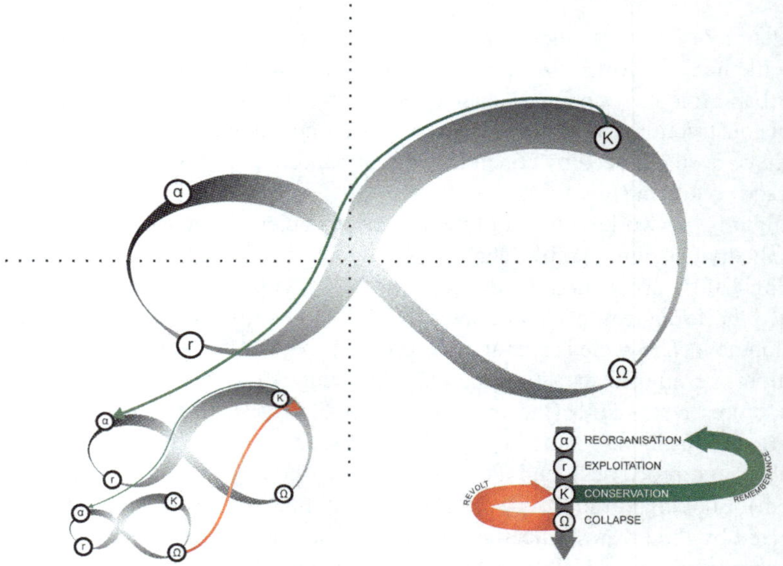

Fig. 8.5 Panarchy model of 2016 studio

concept of space/time hierarchies with a concept of adaptive cycle (Fig. 8.4) and "describes how a healthy system can invent and experiment, benefiting from inventions, that create opportunity while being kept safe from those that destabilize because of their nature or excessive exuberance" (Gunderson and Holling 2009; Holling 2001).

The San Cristobal Panarchy Model

For the 2016 studio, we started with the hypothesis that if we are able to understand appropriate cycles of adaptive change and their scales, we might be able to identify the moments at which to intervene in order to induce positive change.

In our model we have three scales or levels as described by Gunderson and Holling (2009) as the panarchy model. The primary (and middle) level of our model represents the evolution or urban form, the core issue we were examining. The highest level of our model, which cycles slowly, represents the environment. The lowest level, the fastest cycling, represents the role of knowledge in society to inform action. We called this the *noosphere*, a term borrowed from Vernadsky (1945). Our model is not dissimilar to the SES (social-ecological system) described by du Plessis (2008). For du Plessis, the most significant difference between cities and other types of ecosystems is the addition of abstract thought. As Bateson observed:

> The major problems in the world are the results of the difference between how nature works and the way people think. (Bateson 2000)

We have used these models in response to bridge this gap.

This model illustrates how an intervention within the urban form can have an impact on environment and how the intervention is dependent on the "noospheric" level to guide action. Unless there is some positive change within residents and government's frame of mind, unless there is enough encouragement to learn, innovate, and take some risks, unless residents can clearly see benefits of these small disturbances and are equipped to do more, nothing will change, and the whole system will remain in the status quo which has led to the current problems.

The Primary Level: Urban Structure and Typologies

To prepare to identify and model disturbances within the urban structures to improve the environment and the quality of life for the residents and other species in the coupled environments, we collected data on the existing typologies by observation of Puerto Baquerizo Moreno.

As mentioned above, most of the population immigrated to the islands from all parts of Ecuador in search of better quality of life. Although they arrive with different cultures and different preconceptions of a house, most of their dwellings look surprisingly similar (Fig. 8.6).

In describing the typologies, we paid particular attention to degree of their flexibility and adaptability. We observed that boundaries and borders were particularly important and that typologies with more porous elevations appeared to better adapt to changes as compared with typologies that presented hard boundaries by way of fences and walls. Typologies where the private/public interface was ambiguous and therefore able to increase their usable area appeared to perform better economically than those where the public/private interface was clearly defined (Fig. 8.7).

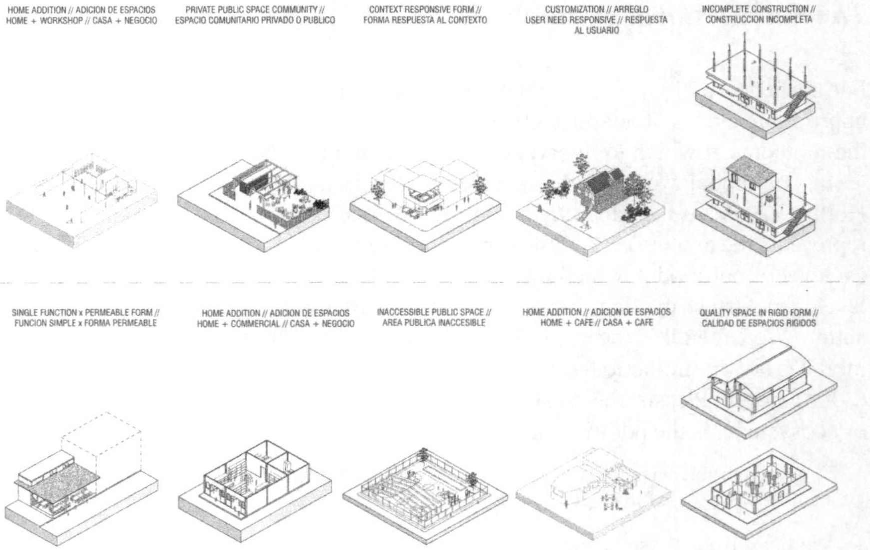

Fig. 8.6 Different typologies

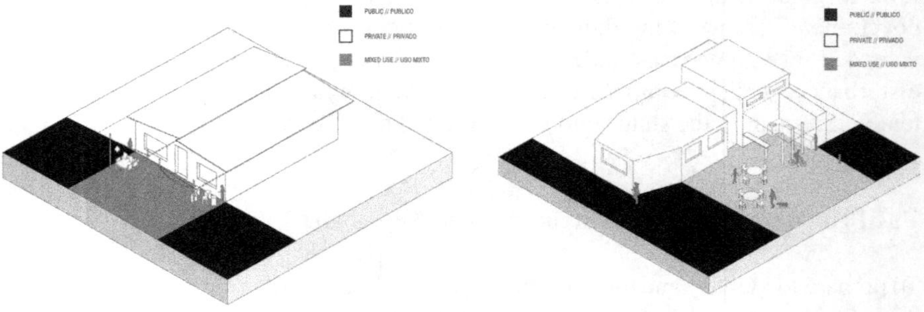

Fig. 8.7 Interplay between public and private (areas in gray color could be calculated twice as private and as public space, therefore substantially increasing the overall area of these typologies)

The most common typology is single-story building with structural elements, such as columns and reinforcement bars, in place on the roof in anticipation of another floor being constructed. We called this an *incomplete* typology. Incomplete typologies were also performing better than completed (resolved) ones. Residents of Puerto Baquerizo Moreno typically enhance their dwellings incrementally, as and when they have enough money to purchase materials or when necessity arises to expand their accommodation (Fig. 8.8).

Most of the typologies located on the three principal thoroughfares, lying parallel to the waterfront, deploy some part of the ground floor for business purposes. In some cases, the business may start as simply a window opened to the street with

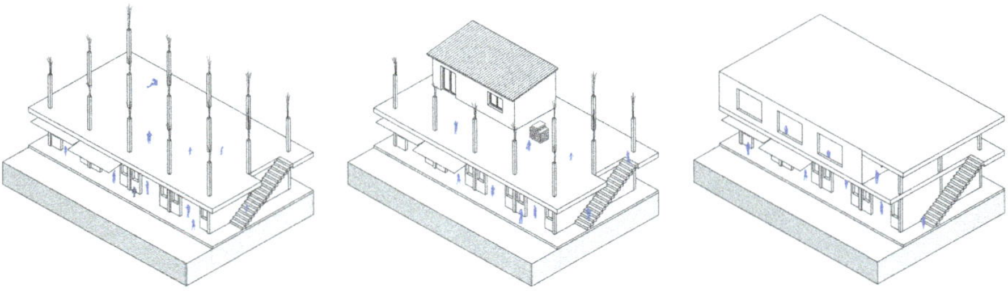

Fig. 8.8 Incomplete typology and ability to expand, grow, and change

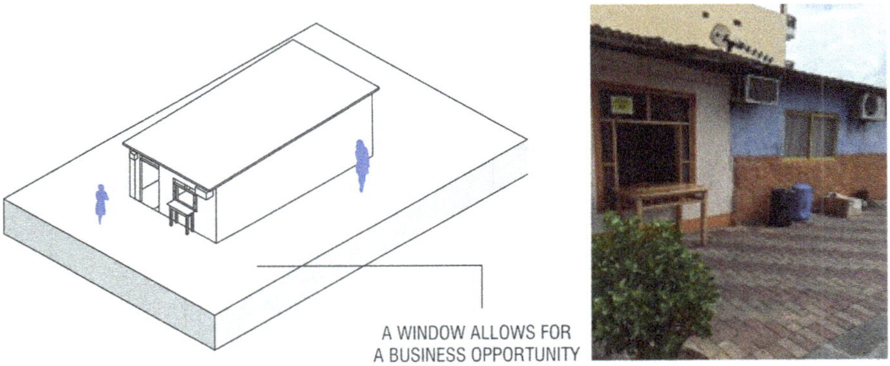

Fig. 8.9 Limited porosity and opportunity for business at the ground level

sign advertising services (such as photocopying) or selling products (such as ice) (Fig. 8.9). Some buildings use their incomplete upper floors as outdoor bars or restaurants. Many of these places are not enclosed, allowing for cross ventilation.

We observed that private enterprise readily expanded into the public domain, placing their merchandise or tables and seats in the public space (Fig. 8.9). This unregulated flexibility between private and public space allows the businesses to multiply their space and generate more profit to then add to their houses and thus to more or additional business opportunities (Fig. 8.7). For example, a bar becomes a restaurant and then a guesthouse as an additional story is added and rooms created.

The inventory recorded indicated that tourism-based businesses became less frequent as we moved further from the waterfront, and therefore dwellings started manifesting different typologies (Fig. 8.10). Some had a small garden in front of the house, surrounded by fence; others elevated building on columns to the upper floor level in order to catch a breeze or a view of the ocean, the ground level being a garage or storage space. The street connection had lost its transactional value and could be passive.

We observed that buildings with commercial facilities, whether on the ground or upper floors, exhibited the capacity to adapt and to expand, whereas buildings that are not accommodating of commercial activities remain the same, often in the same

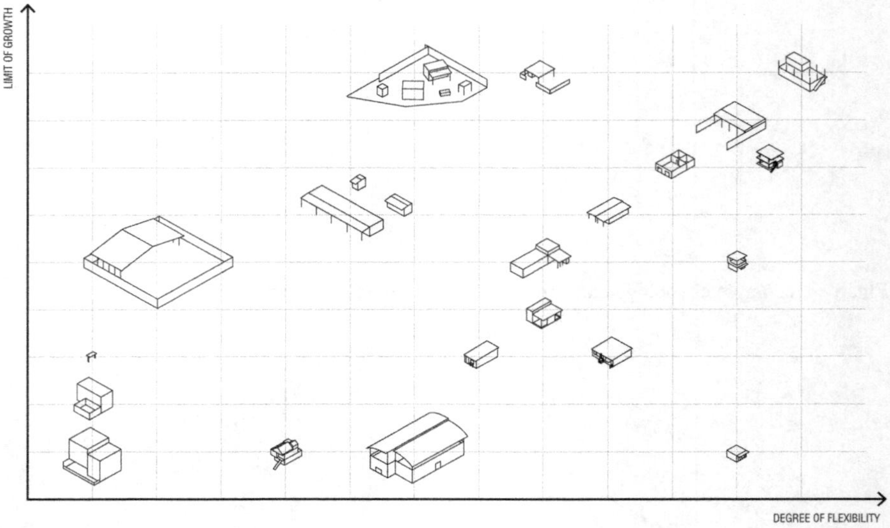

Fig. 8.10 Typologies based on limits of growth and flexibility

uncompleted form for many years (Fig. 8.10). We conclude that it is not only the typology that allows for adaptation and change but also the location and density of pedestrians on the street level that bring added income.

Having catalogued different typologies, we then organized them according to flexibility and limits of growth (Fig. 8.10).

In those structures that allowed for change, we can observe significantly greater permeability to the public faces (Fig. 8.11). In these, the porosity of the elevation allows functions to spill onto the street, only increasing the usable area for commercial or private activities but also encouraging interaction and exchange and supporting economic growth. Conversely, typologies that have barriers surrounding them, fences or walls limiting access to a single entry point, are limited in their ability to adapt and change. We diagramed this to illustrate a desirable state for modification of existing typologies, and new typologies arise from this insight (Fig. 8.11).

The typological catalog does not reveal the extent of abandoned land within the urban area, either overgrown by shrubs, often including invasive species, used for storage or as a rubbish dump (Fig. 8.12). We observed that open spaces by the houses seem to have no value to the residents; there was no visible care or pride related to these areas. We saw no vegetable patches, no cultivated flowers or outdoor furniture. It appeared that family life only happened inside the house. This was very surprising since the Galapagos climate could be conducive to many domestic activities happening outside. Instead the open spaces around the houses looked neglected. This suggested little sense of care for the local environment.

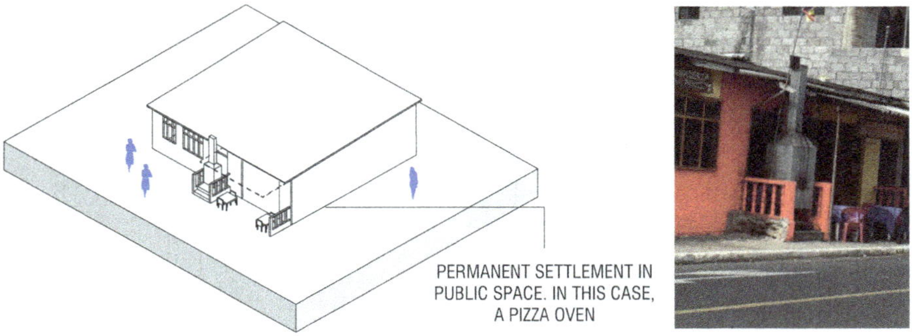

Fig. 8.11 Pizza restaurant expanding into the public domain

The Knowledge Underpinning: The Noosphere

As mentioned above, the lowest- and fastest-changing level in our panarchy model is what we called the *noosphere* (Vernadsky 1945). Vernadsky's proposition is that the noosphere emerges through the evolution and the interactions of human minds. Legal, educational, religious, research, industrial, and technological systems are all part of this noosphere. An important dimension of a resilient society is the richness of knowledge in the society to sustain and evolve within itself. In our model we look at how local knowledge could be built by increasing interaction between the residents. These new interactions will result from the necessity to experiment, innovate, and create new social networks in order to improve their quality of life. This knowledge is important too in broadening an understanding of actions that benefit the environment and contribute to the most valuable assets of Galapagos, its natural habitat.

We differentiate knowledge here into two different kinds, *knowledge of* and *knowledge for*. *Knowledge of* represents descriptive knowledge, how things are, for example, research-generated knowledge. The *knowledge for* is knowledge that can be applied to understand how things might change. This is the knowledge needed by designers, leaders, entrepreneurs, and innovators. In our "noospheric" level, we described too three different types of possible learning: *incremental* (from r to k), *lurching* (from Ω to α), and *transforming* when the learning links one scale of the panarchy model to another (Holling 2001; Gunderson and Holling 2009). This reflects the need to model mind-set change, since without changing mind-sets of the Galapagos residents, government, and tourists, very little could be done to change the existing status quo.

There is evidence that this has happened before on the islands. The people of the Galapagos have had to adapt rapidly to changing circumstances, and their perception of the world and values influencing these perceptions are malleable:

Fig. 8.12 Empty lots within urban area

> As expected, our survey responses revealed that people's productive activity shapes their perceptions, desires, and cognitions with respect to local fauna. Their practical and material concerns determine their belief systems, which depend on whether or not sea lions attract commerce in form of tourism or constrain their livelihood activities, as in the case of sea lions. (Denkinger et al. 2014)

What is evident here is that fishermen did not suddenly become conservationists but rather that they perceived that sea lions had greater value as an attractor for tourism than the threat they posed competing for the fish. This is a situation in which a step is taken toward a sustainable future not driven by a vague existential danger, the threatened future of the planet, but by the recognition that a better quality of life can be achieved.

Quiroga and Ospina (2009) reported that more than 84% of the population thought that science was very important, but besides the study of evolution, it should also be concerned with more immediate topics including migration, public health, and the impact of tourism. There was also a noticeable difference in perception between older and younger generations (Quiroga 2012). Quiroga postulates that this is due to the increased number of educational programs and resources available to conservationists that allow them to spread their message.

It is also evident that people's attitudes are changing and they could change further if the benefits from a sustainable future are better appreciated. The population of the Galapagos Islands is ready for change. The challenge is whether this can be achieved without significant financial support.

It is in this context that the CAS concept of small disturbances in a system can set up tipping points for substantive transitions. The learning is that visible personal benefits can be a crucial component of a larger educational message and that tipping points leading to a better quality of life can also promote a more sustainable and resilient future. In our work, we suggest that these small disturbances for a better quality of life should be placed where they are most visible, within urban environment.

It is also possible that the visible benefits from these small interventions will be able to not only provide new source of employment but create new, Galapagos-specific knowledge available to be exported to the rest of the world (Fig. 8.13).

Fig. 8.13 Possible new typologies

Top Level: Environment

The goal of this design as a thought experiment was to examine how the urban population of the Galapagos Islands can contribute positively in our model of a coupled natural and urban environment. This is highest and slowest changing level of the panarchy model. Most of our interventions were related to minimizing dependence on import from the mainland and building self-sufficient infrastructure.

We have created models in order to visualize how small interventions into the existing urban structure can promote positive changes toward a more sustainable way of life and at the same time improve quality of life with financial benefits for all (residents and government). These models were built on assumption that government may be willing to offer residents microloans that could help to build self-sufficient infrastructure, create variety of new businesses, promote recycling, and increase knowledge and community building (Fig. 8.14).

According to Serres (1992), any healthy system requires constant perturbations that help the system to enrich itself and become more complex, more adaptive, and therefore more resilient. When the system is disturbed and is in the process of reconfiguring, its main objective is to maintain its functionality rather than to preserve particular features or components of the system. It is therefore very important that we have a good understanding of the functionality that we want to maintain and which disturbances or perturbations will not cause irreversible damage to the system even if they are significant.

Opportunities for Interventions and Disturbances

When looking for opportunities for intervention/disturbance, we were also looking for hidden resources which we could use and which will most likely have rapid impacts in environmental improvement. As a first step, we decide to investigate things that were most visibly related to the urban environment that could have both visible and environmental impact and that could also improve the quality of life. Three that emerged were domestic waste, invasive species, and water. Instead of framing them as problems, however, we framed them as opportunities and explored

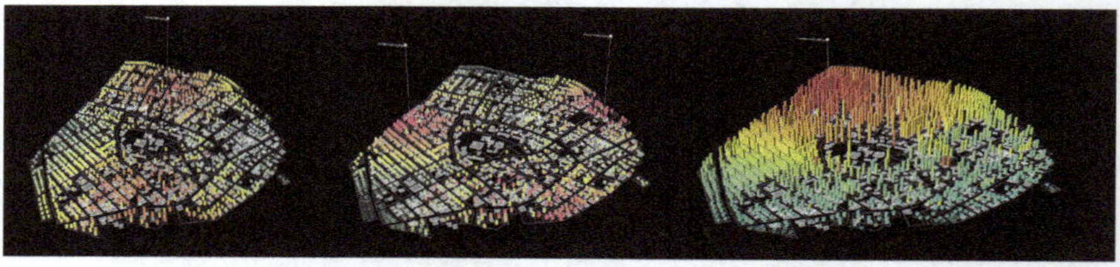

Fig. 8.14 Visualization of impacts created by small, medium, and large interferences into the system

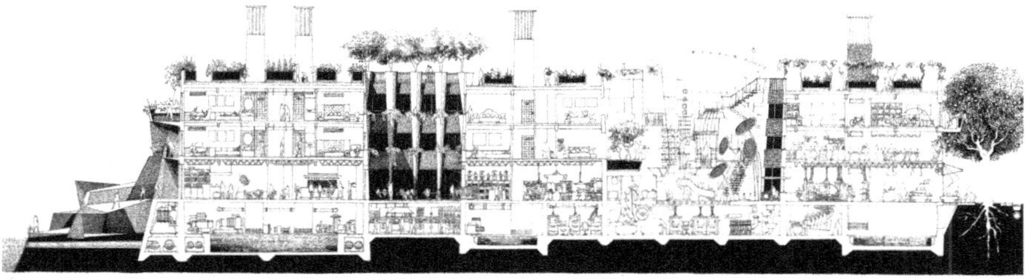

Fig. 8.15 Possible futures

how they might be used as disturbances for positive change and reduce dependency on imports, with all the consequent risks these bring.

Moynihan (1973) describes any types on dependency as an incomplete state of life: normal in the child but abnormal in the adult. This dependency could be compared to immaturity, and as Kant (1970) points out: *immaturity is the inability to use one's own understanding without the guidance of another... Have the courage to use your own understanding*. If residents of the Galapagos could generate their own local knowledge and find opportunities specific to the islands' environment, they may find a path toward self-sufficiency and independence from the mainland for food, energy, building materials, etc. If this local knowledge could also expand toward the creation of new industries, recycling, and local production, the Galapagos may start exporting not only its own products but also their knowledge.

Figure 8.15 provides a summary of our discussion of specific interventions and the ways in which they might be implemented in Puerto Baquerizo Moreno.

Waste as a Resource

Municipalities of the four inhabited islands of Galapagos have identified solid waste disposal as one of the main issues that needed to be resolved (WWF and Toyota 2010).

The problem remains significant, and there are insufficient resources and local expertise to address this problem (Watkins and Cruz 2007). In 2006 outside Puerto Ayora, on Santa Cruz Island, "the Fabricio Valverde Environmental Centre" was established as joint venture between the community and the National Park. The aim was to recycle 40% of waste materials generated on Santa Cruz. A year later in San Cristobal and Floreana, similar initiatives followed, together with environmental education and public awareness programs, which included training and internship for plant technicians (Dunn 2010; Feng 2014; Hardter 2008; Tapia et al. 2009).

More recently, the Galapagos National Park Directorate sought in December 2014 to get rid of plastic shopping bags, of which approximately 4.5 million were used in Galapagos annually, by distributing reusable cloth bags for each family for everyday purchases and charging for plastic bags in shops and supermarkets.

From August 2015 there has been a ban on the use of plastic products (Galapagos.org 2014).

Solid waste on San Cristobal Island amounts to 4.17 tons per day (0.559 kg per person per day). Unfortunately only 10% of this is recycled (Ragazzi et al. (2014). There is no provision to collect scrap metal, computers, white goods, or cell phones. These materials often end up on the rubbish dumps or on abandoned land in the urban area. Even if a proportion of this waste is sent to the mainland, the remaining waste is a significant problem for both land and marine habitats. The need to reduce waste to minimum is very evident, and something much more drastic needs to be done, but not necessarily by government. If residents could see value in their waste, the waste will be able to decrease substantially.

Paper, glass, metal, etc. can all be recycled, but some investigation needs to be done into each product to see if setting up recycling industry on the island will be beneficial or cause more environmental damage. The opportunity is to identify how waste processing can be carried out appropriately locally while also leading to new employment possibilities but also create knowledge specific to the Galapagos Islands, in turn supporting a knowledge-based economy engaged with tourism.

One aspect of this lies in organic waste. On San Cristobal Island, this amounts to as much as 61% of total waste. A large proportion of this waste could be use as fertilizer if there is demand for the product; this could happen if market gardening is established and residents grow food within the urban areas. Abandoned land within urban areas could be replaced by productive landscapes and gardens. The reliance on imported foods will be reduced with access to seasonal, fresh produce, creating employment opportunities and, if endemic produce is cultivated too, supporting another dimension of tourism by way of experiencing local foods.

A substantial part of the organic waste is not compostable as it is the carapaces of shellfish. In particular waste there is potentially another valuable resource, chitosan paste. Chitosan paste has been used as for treating wounds for many years (Ahmadi 2008) but recently has been used for additive manufacturing, including in construction (Oxman 2015).

> We ordered a bunch of shrimp shells, we grinded (sic) them and we produced chitosan paste. By varying chemical concentrations, we were able to achieve a wide array of properties -- from dark, stiff and opaque, to light, soft and transparent. In order to print the structures in large scale, we built a robotically controlled extrusion system with multiple nozzles. The robot would vary material properties on the fly and create these 12-foot-long structures made of a single material, 100 percent recyclable. When the parts are ready, they're left to dry and find a form naturally upon contact with air. So why are we still designing with plastics? The air bubbles that were a byproduct of the printing process were used to contain photosynthetic microorganisms that first appeared on our planet 3.5 billion year ago, as we learned yesterday. Together with our collaborators at Harvard and MIT, we embedded bacteria that were genetically engineered to rapidly capture carbon from the atmosphere and convert it into sugar. For the first time, we were able to generate structures that would seamlessly transition from beam to mesh, and if scaled even larger, to windows… Working with an ancient material, one of the first life forms on the planet, plenty of water and a little bit of synthetic biology, we were able to transform a structure made of shrimp shells into an architecture that behaves like a tree. And here's the best part: for objects designed to biodegrade, put them in the sea, and they will nourish marine life; place them in soil, and they will help grow a tree. (Oxman 2015)

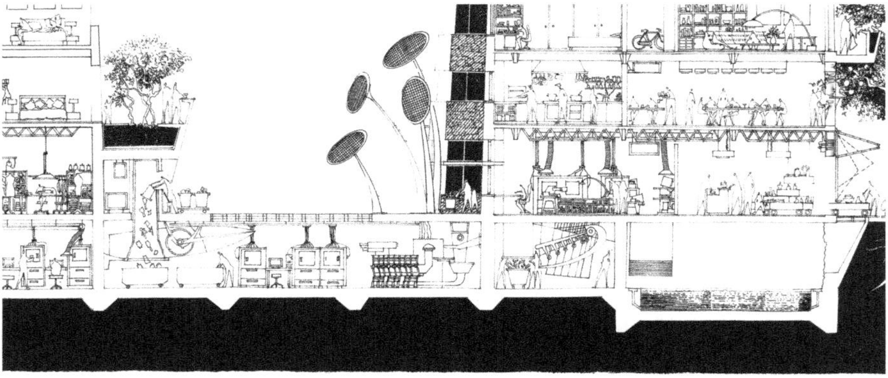

Fig. 8.16 Basement is used to pulverize plastic and shellfish collected from the restaurants. 3D printing labs are located in the basement and on the ground floor next to the shops that sell the products created from recycled materials

While there is much to be done to establish this technology as economically viable for implementation in the Galapagos, it opens our thinking to how additive manufacturing (3D printing) using waste might be a transformative intervention. There are other materials in the accumulated island waste that can be used in a similar manner, plastics being a large proportion of these. With continued advances in waste processing and additive manufacturing technologies, production of filament from recycled plastics or shellfish carapaces will be the norm (Baechler et al. 2013; Fig. 8.16).

Invasive Species as a Resource

One of the most invasive species on the Galapagos Islands is the guava tree (*Psidium guajava*) introduced to the islands around 1910. While there is a native and unique Galapagos guava (*Psidium galapageium*), *Psidium guajava* is highly invasive. Although the attempts are being made to control the spread of *Psidium guajava*, the current techniques are expensive and not effective (Buddenhagen et al. 2004).

Although the guava tree represents a major problem, it has also shown some benefits. Scientists at the Charles Darwin Foundation observed that giant tortoises thrive on young guava trees and avoid native ones (Heisman 2015). Around the world, guava trees' leaves and fruits are used for cosmetic production (Sanda et al. 2011). In Asia guava tree leaves have been used for dengue treatment, and other countries around the globe use guava for inflammation, diabetes, hypertension, pain relief, fever, diarrhea, lung diseases, or ulcer (Gutiérrez et al. 2008). If guava trees can be developed as an alternative construction material, guava trees could become an asset instead of being a problem. Wood from guava threes is resistant to insects and fungal attack; it is heavy and dense and could be successfully used as a building material, for example, as laminated timber, replacing the heavy imported materials.

While there are 11 species of native timber, logging of these has been strictly prohibited since 1995, so introduced trees are the only source and guava the most prolific.

In house construction, the timber is used in complement with imported bricks or concrete or cement blocks. When demolished, the materials are partially reused, but construction waste poses a problem. Bricks and concrete blocks are not recyclable as they cannot be disassembled readily (Jimbo and Grenier 2011).

Previously, the residents of Galapagos only used local materials, mainly lava rock with the timber. The Galapagos Islands generally consist of basaltic rocks and small lava cones from the interaction of plate tectonics and hotspot volcanism (Leonard 2000). Both basaltic and lava rocks are suitable as building materials, but extraction causes impacts, both biological and scenic. The National Park controls the extradition and mines (two on San Cristobal Island) by issuing mining permits and collecting fees; therefore, mining is restricted and expensive.

There is no restriction on excavation of basements and septic tanks for individual houses. A 100 square meter footprint for a basement will produce 300 cubic meters of lava stone that can be then used for construction. Furthermore, a basement carved out of rock can be used as cold storage, rainwater storage, and cooling devices. The rough surface texture of lava rock attracts insect, and insects attract birds; therefore, any structures built out of the lava rock become a living structures, decreasing the divide between built and natural environment.

In this example, we illustrate how turning an invasive species into asset can provide better control of its spread and could provide a reason for more residents to be engage in its control. It could be also the beginning of export of Galapagos techniques and knowledge to the rest of the world through the growth of small industries and related applied research from cosmetics and therapeutic uses to heavier industry producing building materials and furniture. From this arises alternative employment, allowing residents to develop new skills and resilience (Fig. 8.17).

Water

Water has been one of the major challenges in the Galapagos Islands. In 2005 the water management issue was declared as the most urgent problem in Galapagos (Merchant 2008). The only island with freshwater source is San Cristobal, but even San Cristobal has visible substantial problem with water availability and water quality. Most piped water is pumped from subterranean sources, is not potable, and most often is contaminated by leaking septic tanks and seawater. Drinking water is either boiled pipe water or imported bottled water.

Water is readily available. Humidity is relatively high and very constant throughout the year, typically in the 80% (Trueman and d'Ozouville 2010). This is a resource to be harvested by dehumidifying the air with the capability of providing the daily clean and potable water needs for a family of four from a domestic

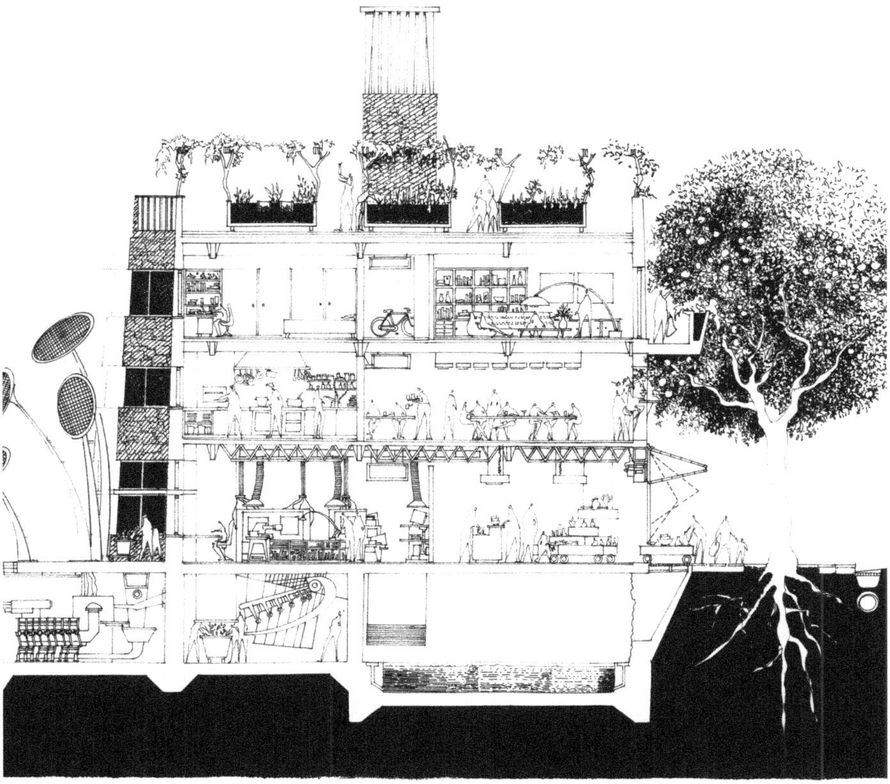

Fig. 8.17 This section illustrates how creating basements could help in obtaining material for main structures and guava trees for laminated beams and traces as well as furniture

installation (Abderrahman 2010; Peters 2016; Klemm et al. 2012). This can be supplemented by catching the mist that is common in the highlands of many of the Galapagos Islands (Chandler 2013). Additional water could be collected from the rain when available (Fig. 8.18).

Summary

In the challenges facing the islands, there are opportunities. In this chapter we have explored one approach to addressing these opportunities. In these, we have demonstrated how small interventions might gradually lead to larger changes within urban structure that realize a more sustainable future. We also illustrated that even the smallest intervention can lead incrementally to the creation of knowledge, skills, and techniques deployable broadly. All of these changes can then change the

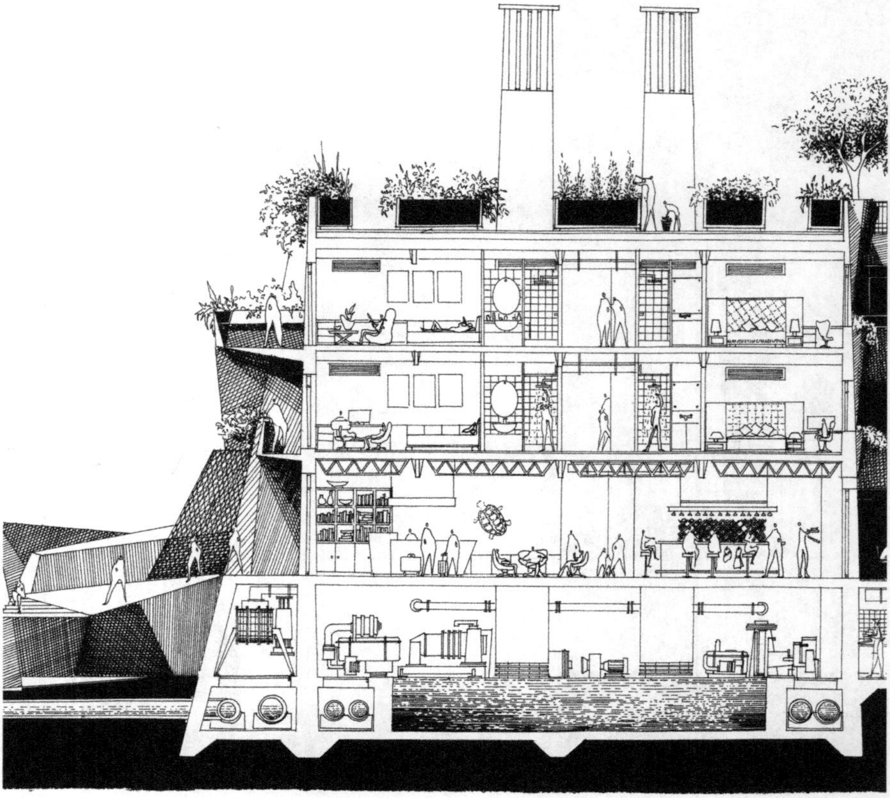

Fig. 8.18 Dehumidifying devices are installed within the wall structure, and the water collected is stored in underground reservoirs that also help to cool the air in the building using wind towers. The collected water is the main water source for the building. Rainwater is used to irrigate urban farming on top of the roof. Seawater is used for toilet flashing, and it also could be used as a heat exchange device

perceptions and behaviors of residents, as has happened previously when they moved from a fishing economy to one based on tourism.

From these incremental urban changes, the larger changes in environment cannot be easily demonstrated, but the consequences can flow with reduced waste, water use, and lower levels of importation using risky transportation. We can therefore have every expectation that environmental improvement will happen as well.

We acknowledge that this exercise is a thought experiment and that the changes are proposed. The model used here is based on the analysis and modelling in the earlier chapters in this book that have traced similar changes so far. In our work, the creation of new employment, new knowledge, and new products might not only decrease dependencies on the mainland but also create export and financial gain not only based on tourism and consumption.

References

Abderrahman WA (2010) Sustainable Water and Food Security and Wastewater Reuse in Saudi Arabia, keynote presentation delivered at a special session on Management Solutions for Water Resources in Saudi Water & Power Forum (SWPF), 5th Oct' 2010, Jeddah, KSA

Ahmadi R (2008) Biocompability and gelation of chitosan-glycerol phosphate hydrogels. J Biomed Mater Res A 86A(3):571–856.

Baechler C, Devuoo M, Pearce JM (2013) Distributed recycling of waste polymer into RepRap feedstock. Rapid Prototyp J 19(2):118–125

Bateson G (2000) Steps to ecology of mind. University of Chicago Press, Chicago

Bensted-Smith R (1999) A biodiversity vision for Galapagos Islands. Charles Darwin Foundation and World Wildlife Fund, Puerto Ayora, Ecuador

Buddenhagen CE, Reteria JL, Gardener M, Wilkinson SR, Soria M, Yanez P, Tye A, Valle R (2004) The control of a highly invasive Tree *Cinchona pubescens* in Galapagos. Weed Technol 18:1194–1202

Chandler D (2013) Fog harvesting: how to get fresh water out of thin air. MIT: Physics Organization. 30 Aug 2013. [Online]. https://phys.org/news/2013-08-fog-harvesting-fresh-thin-air.html

Denkinger J, Quiroga D, Murillo JC (2014) Assessing human-wildlife conflicts and benefits of Galapagos sea Lions on San Cristobal Island, Galapagos. In: Denkinger J, Vinueza L (eds) The Galapagos marine reserve: a dynamic social-ecological system. Springer, New York

du Plessis C (2008) Understanding cities as social-ecological systems. In: World sustainable building conference – SB'08, Melbourne, Australia, 21–25 Sept 2008

Dunn C (2010) 3 things about recycling the US can learn from Galapagos. Treehugger. December

Feng F (2014) Garbage gets a second life in the Galapagos. The Miami Planet

Galapagos.org. Dec 2014 [online]. https://galapagosconservation.org.uk/galapagos-bans-plastic-bags/

Gonzales J A, Montes C, Rodriguez J, Tapia W (2008). Rethinking Galapagos as complex-ecological system: implications for conservation and management. Ecol Soc 13(2):13. [Online] URL: http://www.ecologyandsociety.org/vol13/iss2/art13/

Granda León M, González Camba S, Calvopiña Carvajal V (2013) Measuring poverty in Galapagos. In: Galapagos report 2011–2012. GNPS, GCREG, CDF and GC, Puerto Ayora, Galapagos, Ecuador, pp 84–91

Gunderson LH, Holling CS (2009) Panarchy: understanding transformations in human and natural systems. Island Press, Washington

Gutiérrez RM, Mitchell S, Solis RV (2008) Psidium guajava: a review of its traditional uses, phytochemistry and pharmacology. J Ethnopharmacol 117:1–27. https://doi.org/10.1016/j.jep.2008.01.025. PMID 18353572

Hardter U (2008) Waste management and recycling in the Galapagos. Galapagos Newsletter. December

Heisman R (2015) Galapagos Tortoises Thrive on Invasive Plants: in unusual management twist, researchers learn that non native plants aren't always bad. Earth Island J News World Environ

Holling CS (2001) Understanding the complexity of economic, ecological, and social systems. Ecosystems 4(5):390–405

Jimbo W, Grenier C (2011) The construction sector of Puerto Ayora. In: Galapagos report 2009–2010. GNPS, GCREG, CDF and GC, Puerto Ayora, Galapagos, Ecuador, pp 137–142

Kant I (1970) An answer to the question: what is enlightenment. In: Kant's political writings (trans: Nisbet HB). Cambridge University Press, Cambridge

Karakiewicz J (2016) Interventions in Complex Urban Systems: how to enable modelling to account for disruptive innovation. In: Walloth C, Atun F, Gebetsroither E (eds) Understanding complex urban systems. Spriger Complexity, Switzerland

Karakiewicz J, Burry M, Kvan T (2015) The next city and complex adaptive systems. In: Celani G, Sperling DM, Santos Franco JM (eds) Computer-aided architectural design: next city-

new technologies and future of built environment. Springer, pp 3–20. Spriger Complexity, Switzerland.
Klemm O, Schemenauer RS, Lummerich A, Cereceda P, Marzol V, Corell D, …, Fessehaye GM (2012) Fog as a fresh-water resource: overview and perspectives. Ambio 41(3):221–234. https://doi.org/10.1007/s13280-012-0247-8
Leonard K (2000) The formation and geological setting of the Galápagos Islands. [Online] https://www.geol.umd.edu/~jmerck/galsite/research/projects/leonard/Geospot.ht
Merchant B (2008) Water, water everywhere, but not a drop to drink: the trouble with water in the Galapagos. Treehugger. November
Moynihan DP (1973) Politics of a guaranteed income. Random House, New York
Oxman N (2015) Designing at the intersection of technology and biology. In ted ideas worth spreading. https://www.ted.com/talks/neri_oxman_design_at_the_intersection_of_technology_and_biology?language=en
Peters A (2016) These solar panels pull clean drinking water straight from the air. Fast company: change generation. 19 Oct 2016. https://www.fastcompany.com/3064761/these-solar-panels-pull-clean-drinking-water-straight-from-the-air
Pizzitutti F, Mena CF, Walsh SJ (2014) Modelling tourism in the Galapagos Islands: an agent-based model approach. J Artif Soc Soc Simul 17(1):14 http://jasss.soc.surrey.ac.uk/17/1/14.html
Quiroga D (2013) Changing views of the Galapagos. In: Walsh SJ, Mena CF (eds) Science and conservation in the Galapagos Islands: frameworks and perspective. Springer, New York, pp 23–48
Quiroga D, Ospina P (2009) Percepciones sociales sobre la ciencia y los científicos en Galápagos. In: Tapia et al (eds) Ciencia para la sostenibilidad de Galápagos. Parque Nacional Galápagos, Pt. Ayora, Galapagos, pp 109–126
Ragazzi M, Catellani R, Rada EC, Torretta V, Salazar-Valenzuela X (2014) Management of municipal solid waste in one of the Galapagos Islands. Sustainability 6(12):9080–9095. https://doi.org/10.3390/su6129080
Sanda KA, Grema HA, Geidam YA, Bukar-Kolo YM (2011) Pharmacological aspects of Psidium guajava: an update. Int J Pharmacol 7(3):316–324. https://doi.org/10.3923/ijp.2011.316.324
Serres M (1992) The natural contract. The University of Michigan Press, Ann Arbor
Tapia FJ, Navarrete SA, Castillo M, Menge BA, Castilla JC, Largier J, Wieters EA, Broitman BL, Barth JA (2009) Thermal indices of upwelling effects on inner-shelf habitats. Prog Oceanogr 83:278–287. Elsevier: Science Direct.
Trueman M, d'Ozouville N (2010) Characterizing the Galapagos terrestrial climate in the face of global climate change. Galapagos Res 67:26–37
Vernadsky WI (1945) The biosphere and the noösphere. Am Sci 33(1):1–12
Watkins G, Cruz F (2007) Galapagos at risk: a socioeconomic analysis of the situation in the archipelago. Charles Darwin Foundation, Puerto Ayora, Province of Galapagos
WWF and Toyota (2010) Waste Management Blueprint for the Galápagos Islands. WWF, Puerto Ayora, Galápagos

Chapter 9
Avoiding the Color Gray: Parametrizing CAS to Incorporate Reactive Scripting

Mark Burry, Camilo Cruz, and Geoff Kimm

Introduction and Core Proposition

> By far the most noticeable effect of insular isolation on the birds is also psychological—the extreme tameness of all except the migratory species. Ever since the time of the very first visitors to the archipelago this has been remarked and most human observers have celebrated this fearlessness by knocking as many as possible on the head. (Beebe 1924)

As prior chapters have described, the unique conditions of the Galapagos Islands pose a particular challenge in the interface between the relatively pristine condition of its unique natural environment and the built areas and settlements. Human ecologies that rival the natural ecologies have been insinuated on the islands over several centuries. The human population is expanding and therefore needs to be further accommodated in a way that is not only non-destructive to the existing natural habitat but also seeks to ameliorate the negative aspects of human settlements to date.

The core proposition is necessarily reductive but should not be seen as controversial. Designers receive briefs from their clients: the client expresses their desired outcome from an expert – the professional designer – and evaluates the expert's design proposal. There are arguably infinite spatial configurations for a three-bedroom house, for example, particularly when responding to the unique nature of specific site conditions. Accordingly, the designer does their best to narrow an initially wide set of options to an outcome favored by the client. Often the client might refer their designer to existing properties as a conversation starter, as much as "not like that!" as "these are the features that work for me and these are the deficiencies

M. Burry (✉)
Smart Cities Research Institute, Swinburne Research, Hawthorn, VIC, Australia
e-mail: mburry@swin.edu.au

C. Cruz · G. Kimm
Melbourne School of Design, Faculty of Architecture, Building and Planning, University of Melbourne, Parkville, VIC, Australia

T. Kvan, J. Karakiewicz (eds.), *Urban Galapagos*, Social and Ecological Interactions in the Galapagos Islands, https://doi.org/10.1007/978-3-319-99534-2_9

that would need to be fixed up in the design." The designer will take such precedential input on board, but, essentially, they will commence with a blank sheet with which to "ideate": synthesize and outcome taking the analysis of available relevant data into account.

Given the state of the designers' digital workbenches today, it is not at all clear that designers have made as much effort adopting new media to help their clients become better at briefing them as they have applied to mastering the sophisticated design software at their disposal. Our core proposition is that the emphasis of the design brief preparation can shift from the designer to the client (as much as key stakeholder as end-user), whereby the designer represents the needs and the desires of the client more as skilled agents and interpreters than as masters in the traditional sense: experts whose approach and methodology keep them aloof from the non-cognoscenti. Traditionally design is a top-down approach, whereby the experts exert their expertise from an elevated position. This chapter looks at a bottom-up approach, whereby the client can have a different relationship to the design process. This is not to enfranchise the client as an amateur designer. Rather, it is the quest of ensuring that the person or group who know most about their needs can have more influence in the evolving design. This is explored in more depth later in the chapter.

At the same time as redefining the *client-> architect-> associated professionals* relationship as more of a loop than a line, the designer is also seeking to improve client proxy access to computational design decision support through simple-to-use digital tools. Through employing an iterative satisficing process handling multi-criteria performance goals, complex adaptive systems theory (CAS) can be called upon to handle a large number of competing parameters in an act of synthesizing a brief more attuned to the clients' own analysis of their actual situation and needs. Ultimately CAS can be used to ensure that the brief is flexibly responsive as the design evolves, something that is very difficult to achieve without CAS.

There is already an inherent process complexity for any designer in the traditional role of expert design "master" (auteur). As we trend toward design teams formulated to be more multidisciplinary in composition, invoking end-user (client) human direct inputs *during* the process calls for a fresh look at the computational aspects of digital design. Affording end-users the chance to respond (react), rethink, and change their brief as the design unfolds, renders the computational back-end more complex still.

Parametrizing the design brief and reacting to insights revealed as the computation proceeds offer unorthodox challenges, and our resulting quest seeks to probe four key dilemmas:

1. Can we reconcile quantitative parametrized design inputs with qualitatively motivated shifting inputs as the design brief develops?
2. How do we work with fundamentally competing *sets* of parameters?
3. Is there merit in a bidirectional flow between three distinct information environments: spatial data including 4D temporal representations, 2½ D data matrices and spreadsheets and, crucially, the diagram? (Fig. 9.1).

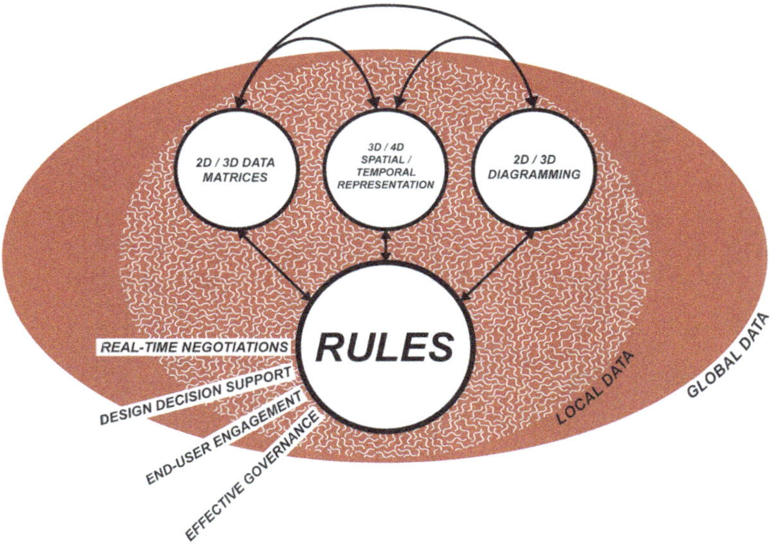

Fig. 9.1 Bidirectional design computation between three linked representation environments

4. How do we avoid zeroing in on a compromise that neutralizes the competing parameters but trends toward a result that lacks any personality – the "gray" outcome?

Toward a Proof of Concept

Taking the particularities of the contentious Galapagos urbanization, i.e., the tension between the preservation of the unique and delicate natural conditions of the archipelago and the development of infrastructure, along with the environmental consequences that it entails, we have sought to develop a proof of concept – a working bidirectional model that allows the designer and end-user alike to input refreshed criteria into an active design computation process knowledgably in the search for a more attuned brief than would traditionally be possible. "Knowledgably" signifies that the design team can react to surprises and unanticipated insights revealed along the way using their various strands of expertise, not least the familiarity of the end-user commissioning the end result, be it a part or a whole building, external environment or the whole urban community. This has been set up as a prototype bidirectional interactive digital design workbench consisting of the spatially rendered model, the multi-leaved spreadsheet and the interactive diagram (Fig. 9.1).

In order to tune this to the October 2016 Galapagos workshop referred to in Chap. 1 of this book, we arbitrarily formulated a set of three indicative and potentially rivaling parameter sets to apply during the workshop:

1. Economic
2. Ecological
3. Experiential (liveability)

By "rival" we mean that the desirable values for one set of parameters might significantly compromise or be compromised by either or both the other two parameter sets – the traded-off satisficed resolutions of which might metaphorically trend to the color gray. Working via an online shared document, we formulated credible parameter sets and individually weighted each parameter as a prioritization; no consensus emerged as might have been expected, which gave the experiment extra credibility.

Desired Outcome

We contrived to establish a workable cluster of three sets of parameters and have a prior sense of the likely difficulty in getting agreement on priorities.

Outcome 1: Consensus around the viability of framing sets of parameters that will work for all end-users.

Outcome 2: A working methodology for subsequent development of the bidirectional flow between information environments that are tailored to particular users of the digital workbench, e.g., "client," "designer," "dweller," "engineer," "policy-maker," "planner," "ecologist," "economist," etc. The outcome was loosely defined as a separate but bidirectionally linked 3D model "information environment," a spreadsheet, and an interactive diagram, each of which can be interactively tinkered with causing a consequential responsive adaptation in the other two information environments. A value could be changed in the spreadsheet, for example, and so the diagram shifts to show the resulting implications visually, and accordingly the 3D model would morph spatially. Of course, there will always be competing choices, pointing us to discover how "hidebound" such a workflow is computationally and to seek the identities of appropriate CAS approaches and "reactive scripts" to work some creative magic invisibly under the hood.

Background

Our proposition is a novel framework to inform decision-making for the process of the generation of architectural form. The idea is to enable designers and/or clients as end-users to cross-reference available information with their own domain knowledge, allowing them to test and reformulate their intentions until they become meaningful for the task, both functionally and aesthetically. The goal is to provide designers and clients with a set of tools to generate architectural artifacts that interact sensibly in a delicate environment such as the Galapagos Islands.

In this chapter we introduce some of the key concepts that inform the development of the framework. In order to do so, a brief description of the process of urbanization in general and how it has developed in the Galapagos is presented, to establish the multiplicity of aspects dealt within the generation of architectural form. A working definition of design is presented next, to clarify its challenges and possibilities, as well as to determine the importance of avoiding the color gray.

General Notes on the Process of Building Human Environments

Being a human-driven process, urbanization is propelled by the needs and desires of a group of people. Unlike what happens with natural evolution, it unfolds as a consequence of a series of decisions taken by groups with different interests and agendas. Seeking shelter, developing social structures for cooperation, and establishing economic activities for subsistence are only a few possible reasons to urbanize a certain area.

To this day urban settlements are considered the embodiment of disorder and chaos, to which the preferred method of control is the imposition of idealized plans. However, for the past 35 years, studies of settlements have unraveled the patterns and structures that lie beneath the apparent chaos, revealing a process of decision-making that enables us to understand more clearly the real complexity of these systems (Batty 2008).

In order to illustrate the complexity of the challenge being addressed, urban settlements can be generally defined as dynamic systems made of buildings and people, constantly traversed by flows of matter-energy that drive their transformations (DeLanda 1997). The most persistent feature of these systems is their physical build, as it constitutes the most rigid and the most durable. Nevertheless, it is the physical manifestation of human settlements what acts as the container that supports the ever-changing interactions of all other non-physical overlaying urban features (Vance 1978). The physical manifestation of urban settlements happens to be the focus of design.

Holland (1995) suggests that human settlements share certain characteristics with ecosystems, which allows us to conceptualize them as complex adaptive systems (CAS) understood as systems lacking in centralized control, never settling into permanent structures and constantly adapting to the new environment. The application of CAS in the way we analyze and conceptualize human settlements has created a major shift in the way we think about urbanized environments.

Human Settlements in the Galapagos Islands

The development of human settlements in the Galapagos Islands can serve as a valuable case study to investigate the shift that the role of designers might profitably undergo from being top-down master planners to becoming expert facilitators of the bottom-up self-organized process.

The Galapagos archipelago is known around the world as a particularly well preserved natural environment, where a high number of endemic species can be found. A less known fact about the islands is that they host a permanent population of around 30,000 people and receive up to 200,000 visitors every year (Gonzalez et al. 2008).

The settling process in Galapagos began with the discovery of the archipelago and developed slowly, focusing on subsistence activities. The promise of more lucrative activities in the late twentieth and early twenty-first centuries triggered a bigger influx of migrants, mainly from continental Ecuador (Walsh et al. 2010), driving the development of support infrastructure and consequently threatening the delicate natural ecosystems present in the islands (Gonzalez et al. 2008; Hoyman and McCall 2013).

This has generated a situation in which the natural environment and human-made systems coexist in tension, where the dispute is one of either idealistically restoring natural systems to their original condition or pragmatically acknowledging the relationship between these competing systems perpetually in conflict and work toward a better integration.

The appeal of the islands as a natural sanctuary has drawn the attention of scientists and naturalists looking to unravel the way in which nature works on the archipelago. When it comes to human action, however, and the development of settlements, the amount of available data is considerably reduced. Studies done in Puerto Villamil show that the complexity of the geography of the area, as well as the regulations that protect certain areas and species, makes urban expansion very difficult. Nevertheless, the environmental impacts of the expansion of human settlements end up being problematic and difficult to address (Walsh et al. 2010).

As it is shown in Chap. 2 of this volume, the urbanization process of Puerto Baquerizo Moreno, located on the island of San Cristobal, has taken place distinctly from the bottom up, as a series of discrete interventions distributed on the territory based on contextual conditions and the regulations in place.

In this chapter we explore the contribution a design approach can make to understanding the development of sustainable human settlements by integrating existing available data into a model that allows designers to interpret, manipulate, and visualize data in three distinct but connected information environments.

Designing Environments

As it is discussed in the previous chapters, the development of human settlements on the Galapagos archipelago has responded to a combination of bottom-up and top-down logics, where apparently the former prevails.

When designers intervene on built environments that behave as self-organizing systems, it is paramount for them to adopt strategies that allow for the accommodation of change and adaptation. Otherwise, design interventions tend to petrify the system into a given state, reducing the ability of the system to adapt, thus becoming an obstacle for its development. To understand how this type of environment develops, from an environmental design perspective (how the built environment comes to be), a complex adaptive systems (CAS) approach is proposed (Batty 2008).

In this section we propose to introduce the use of reactive scripting (Boussinot and Hazard 1996) in the development of digital models for design as it enables the possibility of human input to an otherwise fully automated model, allowing for adaptation and change to become part of the development of the system.

Designing environments for human inhabitation is an open-ended activity that can be conceived as an intentional, constrained, decision-making process where the aim is to transform what exists into what is desired (Alexander 1964; Simon 1969). Many design challenges are characterized by multiple objectives, which are operational representations of the needs and desires of those interested in the transformation of a particular reality. Design options are generally achieved via the production of some sort of artifact, which in the case of architecture and urban design is most likely to be a physical object (Woodbury 1991).

The kind of design discussed within the scope of this chapter falls in a category that Brian Lawson labeled as "environmental design," which encompasses all the disciplines preoccupied with the generation of artifacts that have a practical purpose yet are also conceived to be looked at (Lawson 1990). This classification resonates with what Hillier and Hanson propose regarding the uncertain relationship between the functional and cultural/social dimension of spatial artifacts (Hillier and Hanson 1989). This special characteristic makes environmental design challenges a very particular kind. Often described as "wicked" (Rittel and Webber 1973) or nonroutine (Rosenman 1997), design problems have a set of special properties. Wicked problems can be characterized by the lack of sufficient information to formulate the problem, which leads to the impossibility to determine if the solution is optimal and consequently leaves the generation of something that approximates a solution as the only way to find resolution for such problems, which informs and potentially changes the problem itself. *Wicked* problems are dynamic in nature. On the other hand, nonroutine design problems are characterized by the lack of knowledge regarding the relationship between design requirements and the formal expression needed to satisfy them, which also leads to resolving by approximation. It is possible to say, then, that designers typically have to deal with confusing and incomplete information, which leads to the development of potential solutions that are not necessarily true or false but can only be good or less good.

This is why environmental design tasks are generally addressed using a generate and evaluate synthetic approach (Mitchell 1977), through which potential design options are produced, probed, and evaluated, looking to identify what Simon (1969) refers to as "satisficing" alternatives. In design synthesis, none of the stages can be fully optimized; thus some sort of heuristics and intuition are always involved (Spillers and Newsome 1989). It is possible to claim, then, that environmental design aims to satisfice a range of requirements, rather than to find globally optimal options, especially when dealing with multiple criteria that may be hard to quantify: dynamic – as the set of requirements may change according to time and context – and/or conflicting with each other.

As pointed out previously, an important part of the design process centers on informed decision-making, which renders the task more challenging as the number of constraints, requirements, and desires to be taken into consideration increases. Computational systems are now commonplace in design practice: they are used principally to support decision-making, as they enable designers to generate and manipulate potential option candidates for a particular design challenge, not only by directly interacting with a formal representation of the artifact being developed but also by assigning and changing values to the different attributes that the computer uses to produce the said representation.

The implementation of computational systems for design requires precise definitions regarding the way in which data is represented and interrelated, in order for it to become relevant to the challenge being addressed (Woodbury 2010). Likewise, it is necessary to work from the premise that environmental design deals with wicked problems, which makes it dynamic in the sense that actions taken toward solving a challenge can not only generate new data to be considered as input when further steps in the design process are taken, as they change the context that informs the design itself. It can also change the structure that defines the challenge too (van Leeuwen 1999). To deal with several levels and types of data that are relevant to a design process, as well as with multiple (potentially conflicting) objectives, the approach that is outlined in this chapter and developed as a proof of concept relies on defining the relationships between different representations of the design process, by interconnecting data represented in different ways, as well as on providing modes of interaction between the designer and that data, using methods related to each form of representation.

Traditional environmental design representation uses plans, sections, elevations, 3D models, and diagrams, as these are effective conventions in conveying spatial information, which contains keys to the cultural and social dimensions of the artifact. Performance-based information and functional characteristics, however, are more difficult to understand just by looking at these modes of representation. Conversely, a spreadsheet can contain large amounts of quantitative data, becoming useful when investigating performance-based attributes of a context or a given designed artifact. Extracting qualitative and aesthetic characteristics from them nevertheless becomes near impossible.

In this chapter, we introduce, as proof of concept, an implementation of a third mode of data representation that works by defining relationships between the different inputs available from the databases being linked together. The primary aim of this model is to constitute the missing link between a purely data-driven representation and the common tools for the development of architectural form, as it enables the exploration of relationships between the two modes of representation that have already been described. We do this by incorporating what Boussinot and Hazard (1996) defined as "reactive scripting" to the implementation of this proof of concept, since it enables an interactive process between the computer-based model and the user.

Reactive scripting defines an event-driven procedure, where an interpreter can react to several commands in parallel (Boussinot and Hazard 1996). Originally developed in computer science, reactive scripting lends itself to be applied in a design environment, as it constitutes a method to build and accommodate complex behaviors. Design applications can be found in the interaction between computerized processes and human intuition: where the computer finds more than one way of resolving a given operation, the designer is prompted to select which one is the more adequate for the task, using criteria that might not necessarily lend themselves to computational processes, notably aesthetic and other qualitative considerations. This makes the development of computerized frameworks less intensive, in the sense that it is not necessary for the developer to anticipate every potential challenge nor option space: the interaction between digital media and human user becomes, of itself, a complex adaptive system (CAS). Design inspiration continues to occur during design development, will not necessarily be logical and is often most inconvenient but where possible will require accommodation and a certain amount of rework.

This proof of concept offers a spider chart to the user. In the diagram, radii lines represent attributes, the values of which are illustrated as a point on each radius line and indicate the influence of each attribute on the overall output (refer to Fig. 9.5 – left). The value points are connected by a boundary line to form a polygon, in order to make the relationships between attributes more explicit. The goal is to facilitate the discussion, negotiation, and decision-making, for the actors involved in the process. To achieve this, we investigated the development of visual representations of the effects of their design requirements – advantages and trade-offs.

Proof of Concept Implementation

In this section we explain the actions taken toward the development of a tool that make trade-offs explicit to the tool user, providing them with direct feedback of actions they have taken.

Algorithm Overview

The proof-of-concept computational design tool implements a continuous, real-time reactive scripting system that allows a designer to specify representational parameters described by a dimensionless numerical value. By assigning weighted relationships between those parameters, the parameter values based on the current state of the system are updated iteratively. In addition, the designer must be able to specify one or more parameters as primary or independent to allow them to control those parameters' values manually and hence adjust the overall state of the system.

The Grasshopper™ visual scripting environment in the Rhinoceros 3D™ modeling application was chosen for its ubiquity, suitability for rapid prototyping of tools, and support for graphical display of data.

Managing Complexity for Design Input

Conventionally, the building "designer" is a professional skilled in listening as the client express their desires, distilling their desires and expectations as a formal set of requirements expressed through a "brief," and ultimately interpreting the requirements as a "program" for a building. The completed building, when it emerges, is a spatial interpretation of the client's desires.

The design, as it evolves, requires feedback from the client who will be invited to review progress in a series of meetings, and it is usual for the existing parameters to be revised along with the introduction of fresh parameters that only occur as a result of the review of the design as it emerges. This iterative process constantly revisiting previous decisions is expensive in both time and the continuous revision of work-to-date. Computation and the emerging possibility of real-time feedback in scenario tests offer a different kind of client-designer relationship, one where "what ifs?" can be explored together on the spot rather than through a sequence of separate meetings. Effectively, the client is brought more actively into the design loop as a bottom-up participant. Our proof of concept is therefore motivated by the opportunity computation offers in this regard: to review this aspect of the fundamentals of building design and procurement.

What Makes the Design Process a Complex System?

Unfortunately, while there is much logic in design as a process, design, as an act of synthesis based on the analysis of diverse sets of data, cannot be argued as being, in itself, logical. For instance, it is inherent in the design challenge that the relationships between parameters (data-driven) might not be clearly understood nor, at the outset, even defined at all. Indeed, at the outset of a design and possibly well into the experimentation stage, which parameters should be included or excluded might even be unknown. Hence, a core issue in the script development is handling a large number of competing parameters in a way that allows a designer and client together to add, remove, and adjust parameters elegantly and manipulate the connections between them in real time. A development priority of the tool therefore was to manage user interface complexity to allow natural, uninhibited, and meaningful exploration of CAS design spaces through intuitive manipulation of parameter sets and combinations in a way that does not rely on design expertise.

In creating this design tool, three experiments were undertaken. Two initial, tabular approaches to managing the complexity of the system as presented to the designer in the user interface were prototyped: first, a 2D array of Grasshopper Control Knob components and, second, a tabular layout using the *HumanUI* Grasshopper add-on. Both approaches failed to sufficiently manage the complexity but are worth mentioning as counter examples and useful failed prototypes leading to the final approach.

First is the early test of a 2D array of Control Knob Grasshopper components (Fig. 9.2). This approach used a grid of Control Knob components to represent and weight connections between parameters. In this configuration, it was impossible to easily update with new parameters any system of more than a few parameters due to the requirement explicitly to manage connecting "data" wires and components in the Grasshopper document. To add or remove a parameter in a system with n parameters requires approximately $3n$ manual alterations to the Grasshopper script; this would rapidly become burdensome to the designer and disruptive to the design workflow.

Second is the early test of text-defined parameters and a tabular user interface created with the HumanUI Grasshopper add-on (Fig. 9.3). This approach sought to improve on the previous one by allowing parameters and their connections to be defined in a text table that was then parsed to create, through HumanUI, a popup user interface. In the interface, those connections defined in the text table as having a weight of zero were grayed out to mark them as redundant. While the process of adding or removing parameters was simpler than in the preceding approach, it was still onerous to the designer as it required him or her to explicitly create and connect a number slider component in the Grasshopper script for each parameter and maintaining the text table is prone to error and not intuitive.

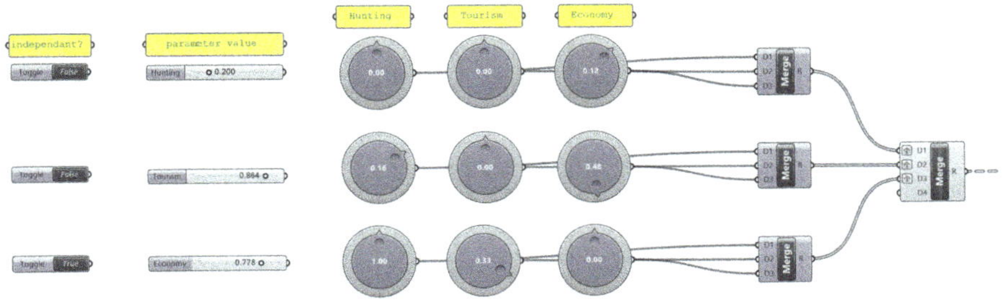

Fig. 9.2 The 2D array of Control Knob Grasshopper components early approach

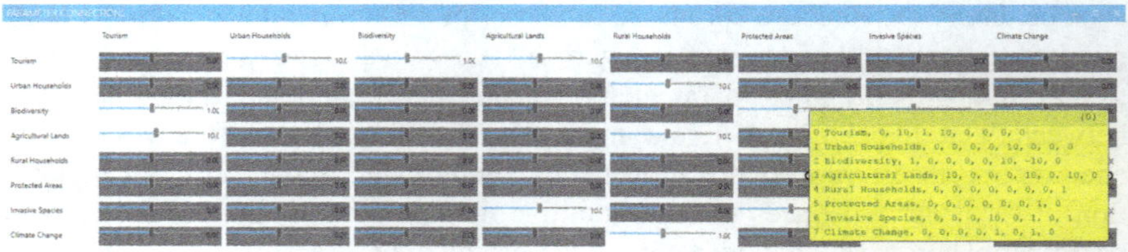

Fig. 9.3 The tabular user interface early approach with superimposed text-defined parameters

Both these approaches had additional flaws: the visual legibility of connections within parameters is "crowded out" by a rigid layout that does not reflect the actual relationship structure and the visual preponderance of irrelevant parameter connections. The simplest example of the latter is that no parameter in the design scenario has a direct connection to itself, yet a connection is still displayed in the user interface or that parameter connections with no weighting and hence redundant are still displayed. It was clear that an approach was needed, whereby only the connections that are explicitly defined are displayed to the designer and in which the relationships are shown as part of the Grasshopper script structure.

Object-Oriented Approach

Reactive scripting cannot truly work if irrelevant data obscures the relationships present and the decisions that a designer must make. An approach using encapsulation of the design scenario state allows designers to work with only those parameter connections that are relevant. The approach works by treating Grasshopper component connecting wires as conduits for the flow through the script of custom data objects that fully describe either a parameter or a parameter connection. Hence, instead of enumerating all possible connections even if not used, as occurred in the tabular approaches, only those connections that are chosen by the designer are present in the script and displayed in the user interface, and the actual structure of the parameter relationships is reflected in the visual script (Fig. 9.3).

A Grasshopper add-on was developed in the C# programming language to provide three custom components:

1. A *parameter* component to create a parameter (Fig. 9.4 [red]). The component accepts inputs of a Boolean value to indicate if the parameter is independent and a number slider with range 0.0 to 1.0 to give the current value of the parameter. The component outputs a single custom parameter data type value containing the assigned information including the name of the component as given by the designer through the standard Grasshopper user interface.
2. A *connection* component to create and weight a possible parameter connection (Fig. 9.4 [green]). The component accepts a single input of the parameter it is to weight; each parameter component in the script may connect to zero or more

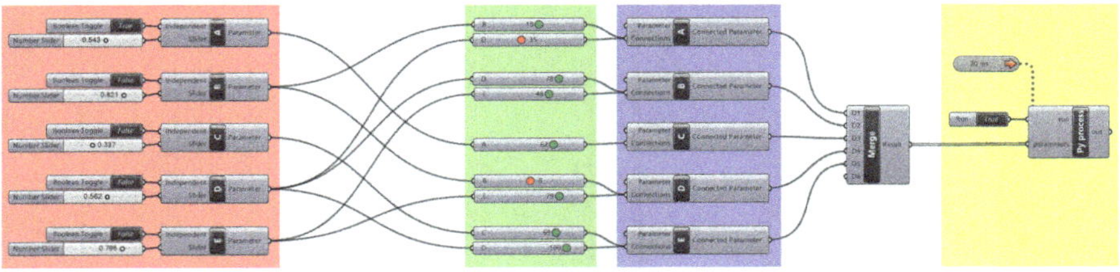

Fig. 9.4 The object-oriented approach showing parameter components (red), connection components (green), connected parameter components (blue), and the Python script (yellow)

connection components. The component exposes a custom number slider graphical user interface in the range −100.0 to 100.0 to allow a connection to be weighted. A weighting of 0 is not allowed in the interface as any such weighted connection is redundant and must be removed by the designer to preserve the visual clarity of the design scenario state. For additional descriptiveness, the number slider control point displays in red for negative numbers and in green for positive numbers. The component outputs a single custom connection data type value containing a parameter identifier and the assigned connection weight.

3. A *connected parameter* component to connect to a parameter one or more other parameters (Fig. 9.4 [blue]). The component accepts inputs of a single parameter value to specify the parameter to which to connect and one or more connection values to establish the weighted relationship with other parameters. The component does not allow a parameter to be connected to itself. The component outputs a single parameter value, corresponding to the specified parameter input, in which the weighted connections are enumerated.

A script in a Grasshopper Python™ component that accepts all *connected parameter* outputs is used to calculate new values for the number sliders input to the *parameter* components (Fig. 9.4 [yellow]). In this script, each new slider value for a parameter is the normalized summation of the weighted current slider values of connected parameters. The reactive scripting computational loop is achieved by the Python script running on a Grasshopper timer, here set to 20 ms and direct programmatic access to the Grasshopper document to update the number slider values of all dependent parameters.

This object-oriented approach is extensible to other user interface environments in addition to Grasshopper and allows a user (designer with client) to focus in real time on only those relationships assessed as being relevant to a design scenario and to easily add or remove parameters and connections.

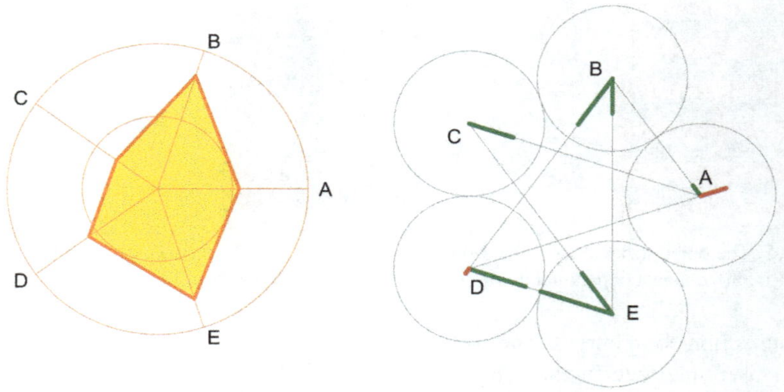

Fig. 9.5 Spider chart (left) and parameter connection graph (right)

State Display to the User

Two methods are used to graphically display to the user the state of the reactive script: the spider chart and a *parameter connection graph*.

The spider chart presents the current values of each parameter in the system, as described elsewhere in this chapter (Fig. 9.5 [left]). The central point represents the minimum possible value, and two concentric circles represent the middle and maximum possible values.

An additional graphical output for the designer is the parameter connection graph that displays all current connections along with the weights assigned to those connections (Fig. 9.5 [right]). For each parameter, each connection it has to another parameter is shown as a gray line over which is displayed a colored line of length proportional to the current weight for that connection; either green and toward the other parameter for a positive weighing or red and away for a negative weighting. The graph provides an alternative to viewing relationships in the Grasshopper script structure itself and one that can be more intuitive for a lay-audience.

Discussion

The proof of concept implementation was tested during the workshop held in Puerto Baquerizo Moreno (Island of San Cristobal), in October 2016. The aim was to explore ways in which this tool could be useful for illustrating the relationships between different attributes of the Galapagos environment, to inform decision-making for the development of human environments. Experts representing a variety of disciplines were able to find common ground around issues regarding tourism, built environments, economic incentives, agriculture, and the ways these are interrelated, by interactively generating diagrams to illustrate these relationships. This encourages us to continue the development of this tool, as it appears to represent a viable alternative to establish a link between the traditional design process, driven by the intuitive manipulation of spatial diagrams (3D models), and a data-based approach that relies on empirical quantitative available data (spreadsheets).

The reactive scripting approach given in this chapter presents a workflow that can be complementary throughout the design process. It can be first deployed to navigate an intractable design scenario and generate community consensus and engagement and harness a community's knowledge and assumptions and then used as a tool that embodies the outcome of that process and allows decisions to be made and feedback on a developing design response in the established option space to be given by non-designers. The proof of concept demonstrated the flexibility between three modes of data representation and that of the system to not only adapt to new parameters during the design process but also accommodate entirely new parameters and connections between them, well after the design process has initiated.

This approach, as tested with the proof of concept implementation in the workshop, can act as a facilitator to generate consensus and a better understanding of a design challenge through discussion of what parameters are important and what connections and trade-offs may exist between them. In this case, reactive scripting is better suited to more nebulous issues dealing with abstract relationships, such as those involving the interaction within and between nondeterministic systems: the environment, for instance, or the economy, or human impulses. When used for these systems tending toward the more abstract, the parameter values might be dimensionless, that is, have no units, as used in the workshop, and the import of those values has to rely on the conceptual framework embodied in the minds of the tool users.

Reactive scripting can also be used to formalize a system of predefined parameters and relationships to allow users to see what trade-offs and sacrifices might need to be made to achieve outcomes they considered desirable. The bounds and mechanism of this option space can themselves be informed using reactive scripting as discussed in the previous paragraph. Here, reactive scripting is more apt for situations where there are more defined relations between concrete parameters, such as if it is used for generating a house plan. When used for these more practical situations, the parameter values might have well-defined meanings; parameters could be expressed in a common unit so that meaningful parameter values and connections can be assigned. For example, in a reactive scripting tool for a house plan, the common unit might be financing to meet a budget, or time to meet a schedule. Where multiple criteria are needed, such as in a reactive script that allows a house plan to satisfy both budget and time considerations, conversion formulae may be used to allow cross-metric weightings so that two or more metrics might be used in the same reactive script.

A significant feature of both modes is that they may facilitate the addressing of instinctively understood concerns that might not be formally defined. Some concepts captured in the tool might not be directly stated but might emerge instead from the understanding of the user and designer. That the concept exists and is modeled in the tool might be understood explicitly, but the actual criteria might be subject to an implicit or hazy understanding. An example of this is the use of the tool to generate a house plan: the tool might function here in terms of cost, such as a user who wants more rooms might need to use a less expensive cladding, but the idea of community status would be clearly understood by the user and designer even though not explicitly displayed in the tool. Reactive scripting here assists the designer to satisfy pragmatic constraints while considering intangibles such as cultural and societal imperatives that only emerge as the design advances.

There is no firm division between the two described modes of reactive scripting, and the flexibility to segue from the conceptual to pragmatic in order to incorporate and formalize intangible considerations as a design challenge exploration progresses has two particular implications. First, the very nature of a solution to a design challenge might be unknown at the outset of the design process; although there might initially be an assumed avenue of response, the ultimate solution might be different in kind, not merely by degree. The reactive scripting method allows a designer to model a design challenge and response with increasing concreteness from early assumptions. Second, reactive scripting presents a method to generate a design options space uninhibited by inherent encultured meaning. Although the reactive scripting framework may be well-defined, such as the interface presented using Grasshopper in the proof of concept implementation, the negotiation of what parameters are important and how those parameters are linked is generated in response to the decisions of the designer guided by real-time collaboration with an individual client or a wider community. Here, reactive scripting is a valuable tool for developing a design response in situations in which there is possibility of cross-cultural communication issues, cultural misunderstandings or assumptions on the part of the designer due to his or her native world view not necessarily coinciding with that of the client.

A further implementation of the reactive scripting approach should present a decision event to the designer when there is more than one possible resolution for updating the co-dependent parameters. The proof of concept implementation demonstrates the existence of and need to address this not unusual case of multiple resolutions. Whereas the system and its representation in the spider chart is usually stable, it does fall into a paroxysmic state when objectives compete without a resolution. The current tool handles this to a degree by allowing the designer to moderate the state of the system by adjusting parameter weights (values) and relationships but does not expressly inform him or her which parameters are in conflict, nor what possible resolutions exist. The solution in a further iteration of the tool would be to prompt the user to specify the alternative route to follow.

Conclusion

> When the problems of migration, of establishment and of adaptation to life on these islands have been solved, we will know much more about corresponding conditions not only in mainland jungles but in the turmoil of Broadway and Regent Streets. (Beebe 1924)

Whereas the traditional role of designer as master (of the design process) is predicated on notions of top-down professionality, from the client's perspective, the designer is working remotely, magically producing evolving design iterations presented to the client from time to time. The process trends down toward a concomitant reduction of complexity with an optimal outcome being the goal – simplexification. This is one characterization of the design process. The designer's particular *modus operandi* is typically hidden from the client's view, and, to a certain extent, the client explicitly has to trust the designer's judgment.

By bringing the client actively into the design process with a bottom-up approach, the design as a synthesis dealing with complexities otherwise invisible to the client becomes overt and can be instrumentally accommodated as a "design tool." In so doing, the tool formally admits human agency as an unpredictable component of the adapting influences within the design process, construed in this way as a CAS. The intuition and insights of the designer will be on view to the client, respected but unlikely to be shared.

Design may include problem-solving as part of its zeroing in on options, but it is not problem-solving per se. Design is not a program and is therefore difficult to program in the computer software sense. Yet design is uncontestably a system in the sense that it is a set of principles and procedures framing a defined action for a specific purpose. In this chapter we have alluded to design being a preternaturally complex process, one starting simply with a metaphorically blank sheet of paper ideating toward a successful synthesis informed by or based on sets of parameters some of which will only enter mid-process. Design is a complex system. For design to be understood as a complex adaptive system, we have had to look closely at what "adaptive" might mean in this context.

As a crucial component of CAS, "adaptation" begs questions around the nature of agency when thinking of design as a CAS. Would there be any difference between a complex adaptive system and a complex *self-adaptive* system? This would depend on which side of the design-as-methodology fence one resides (as opposed to design-as-intuition). Despite there being no evidence (yet) of a sophisticated building being derived entirely using computation with a set of parameters and data complete at the outset, this is still regarded by many as a Holy Grail. Adaptivity that affords variation and surprise within CAS that does not involve human agency would need to be classed as self-adaptive. It is possible that outside the design-as-wicked-problem world, adaptive would implicitly be self-adaptive. In this chapter, we have shown that when design is treated as a CAS, human agency is still required as part of the adaptive strategy to respond effectively to surprising new insights and fresh opportunities not presented at the outset. The tools we developed to provide proof of concept were necessarily modest in nature but ambitious in terms of contribution to the design computation debate. The Galapagos presented an intriguing set of paradoxes to work with and a perfect sandpit for wrestling with the task of taking design deeper into the CAS realm and enfranchising the client as end-user more effectively in the design process along the way.

References

Alexander C (1964) Notes on the synthesis of form, vol 5. Harvard University Press, Cambridge, MA

Batty M (2008) Cities as complex systems: scaling, interaction, networks, dynamics and urban morphologies. In: Encyclopedia of complexity and systems science. Springer, New York, pp 1041–1071

Beebe W (1924) Galapagos world's end. G. P. Putnam's Sons, New York/London

Boussinot F, Hazard L (1996) Reactive scripts. In: Real-time computing systems and applications, 1996. Proceedings, third international workshop on real-time computing systems and applications. IEEE, New York, pp 270–277

DeLanda M (1997) A thousand years of nonlinear history. Zone Books, New York

Gonzalez J et al (2008) Rethinking the Galapagos Islands as a complex social-ecological system: implications for conservation and management. Ecology and Society 13(2): 13. [online] https://www.ecologyandsociety.org/vol13/iss2/art13/

Hillier B, Hanson J (1989) The social logic of space. Cambridge University Press, Cambridge

Holland JH (1995) Hidden order: how adaptation builds complexity. Basic Books, New York

Hoyman MM, McCall JR (2013) Is there trouble in paradise? The perspectives of Galapagos community leaders on managing economic development and environmental conservation through ecotourism policies and the Special Law of 1998. Journal of Ecotourism 12(1):33–48

Lawson B (1990) How designers think: the design process demystified, 2nd edn. Routledge, London/New York

Mitchell WJ (1977) Computer-aided architectural design. Wiley, New York

Rittel HWJ, Webber MM (1973) Dilemmas in a general theory of planning. Policy Sciences 4(2):155–169

Rosenman MA (1997) An exploration into evolutionary models for non-routine design. Artificial Intelligence in Engineering 11(3):287–293

Simon HA (1969) The sciences of the artificial. The MIT Press, Cambridge, MA

Spillers WR, Newsome SL (1989) Design theory: a model for conceptual design. In: Design theory '88. Springer, New York, pp 198–215

van Leeuwen JP (1999) Modelling architectural design information by features: an approach to dynamic product modelling for application in architectural design. Technische Universiteit Eindhoven, Eindhoven. https://doi.org/10.6100/IR524571

Vance JE (1978) Geography and the study of cities. American Behavioral Scientist 22(1):131–149

Walsh SJ et al (2010) Community expansion and infrastructure development: implications for human health and environmental quality in the Galapagos Islands of Ecuador. J Lat Am Geogr 9(3):137–159

Woodbury RF (1991) Searching for designs: paradigm and practice. Building and Environment 26(1):61–73

Woodbury RF (2010) Elements of parametric design. Routledge, London/New York

Chapter 10
Critical Paths to Sustainability: The Research Challenge from Island Urban Systems

Luís M. A. Bettencourt

A Reflection and Outlook

Evolution is special in islands. During his visit to the Galapagos in 1835 (Darwin et al. 1989), Darwin must have realized that island ecosystems constitute a set of independent experiments in how biological species develop as a result of their adaptation to new environments and associated forces of natural selection. From then till today, many critical insights into the nature and dynamics of biogeography and evolution are due to the study of island ecosystems (MacArthur and Wilson 2001; Whittaker and Fernández-Palacios 2007).

Evolution, however, proceeds not only by natural selection but, indeed, as the result of selection and shocks of any kind. The most compelling evidence for biological selection at work comes arguably from artificial selection – such as the breeding of dogs, cattle, or maize, by means of which humans shape these species for specific looks or purposes (Darwin and Mayr 2003). At present, most environments on Earth are disturbed by human activities (Crutzen 2006; Ellis and Ramankutty 2008; Ellis 2011), and while this creates sheer destruction in many places, the more complex reality is that nature continues to adapt and change in response to human activity; the islands of the Galapagos are no exception; see Chap. 6.

Thus, human activities create not only habitat loss and degradation directly but also a number of selective forces to which ecosystems respond and, given enough time, adapt. This adaptation shifts the emphasis (Chap. 4) for creating sustainable development away from simple conservation and places it on the forces created by humans on ecosystems. Critically, this shift of perspective leads to a more dynamic

---

L. M. A. Bettencourt (✉)
Mansueto Institute for Urban Innovation, Department of Ecology and Evolution, Department of Sociology, University of Chicago, Chicago, IL, USA

Santa Fe Institute, Santa Fe, NM, USA
e-mail: bettencourt@uchicago.edu

T. Kvan, J. Karakiewicz (eds.), *Urban Galapagos*, Social and Ecological Interactions in the Galapagos Islands, https://doi.org/10.1007/978-3-319-99534-2_10

view of coupled urban-natural systems (Chap. 3) away from static conservation. Importantly, it opens up the possibility for future co-development and mutually beneficial synergies between nature and human societies.

Human development in islands is also special. Processes of economic growth, for example, though ubiquitous around the world today, still remain poorly understood. Policies to create economic development, for example, are variable in their attainments, and local successes remain largely not replicable when transplanted to new contexts. On a broad scale, we know that additional energy and resources are typically necessary for economic growth to persist. This view is often overextended and argued to imply that any new economic growth will require large amounts of energy and other resources typical of previous growth associated, e.g., with manufacturing. This is not true because most of the highest value-added goods and services are informational and urban, so that higher-income economies can indeed be less energy intensive in terms of unit of GDP per unit of energy used (Bettencourt and West 2010).

Nevertheless, this nexus between energy consumption and the creation of goods and services, which ultimately follows from the second law of thermodynamics in any complex (dissipative) system, creates pressures that in conventional scenarios make economic growth unsustainable because nonrenewable resources are consumed and not recycled within human societies. Thus, inputs to human societies from natural systems are necessary, for example, as timber or fossil fuels, while outputs as wastes are dumped back onto natural environments, such as CO2 or dirty water. These resource flows constitute, in more specific terms, drivers created by human societies on ecosystems, which may or may not be destructive, depending on what they are and how they are structured.

Additional perspectives on the drivers of economic growth emphasize knowledge and technological innovation as necessary for sustained growth (Romer 1990; Barro and Sala-i-Martin 2003). Other approaches to economic growth emphasize "better" institutions that reduce transaction costs (efficient bureaucracies, low corruption, stability) and create incentives for innovation and human capital accumulation (North 1987). Collective institutions, particularly those dedicated to handling commons, are also essential for responsible and sustainable stewardship of natural resources (Acemoglu et al. 2005; Jones and Romer 2010).

What this body of knowledge tells us is that *how economic development happens* matters a lot for sustainability. Though many past and present patterns of economic activity are unsustainable to natural environments, there are almost certainly paths to be discovered, associated with greater prosperity and well-being for both humans and nature; see on particular Galapagos narrative in Chap. 5.

Identifying such paths to sustainability requires active and intentional research in specific systems. The key property of islands is that flows in and out are much more clearly defined and measurable. The Galapagos, in particular, already have a number of systems in place to keep track of several of these key flows, such as human migration to the islands, boat movements, and tourism. Trade flows between the islands and the continent can also be tracked, measuring imported quantities such as food, building materials, vehicles, energy products, and other consumables. One aspect is addressed in Chap. 7.

Detailed studies of these flows and their transformations in the several island towns have the potential for providing a full, dynamical characterization of how the human presence in islands functions, how it creates rising livelihoods and value, and how its adverse effects can be mitigated or eliminated toward creating truly sustainable development; see Chaps. 5, 6, and 7.

A Research Agenda for Sustainability in Island Systems

The general considerations of the previous section map into a general scheme, organizing necessary research toward a sustainability plan for any island system:

A research agenda can be formulated in three parts:

1. **Map the structure of inflows and outflows to each system**. This includes energy, resources, people, construction materials, etc., as inputs, and money, wastes, and people as outputs. This produces an *input-output matrix*, similar to what is sometimes done in both regional science (Isard 1972) and urban metabolism (Kennedy et al. 2011; Sahely et al. 2003). Critically, this methodology unifies these two approaches by tracking and connecting interdependent flows with different units, such as people, energy and goods, and their economic value.
2. **Create a model of each human settlement to track the local transformation and accumulation of inputs**. This includes the consumption of certain products, the expenditure of energy, and the creation of corresponding wastes, as well as the accumulation of people and construction materials. This model should be spatially explicit and disaggregated, so that construction, population, and economic activity can be understood in detail.
3. **Propose and evaluate explicit interdependent changes to inputs, transformation, and outputs toward sustainability**. This can happen, for example, through the substitution of inputs and/or the modification of processes, such that adverse effects can be eliminated or made beneficial while desirable outputs can continue to grow (Jacobs 1985). This means, for example, substituting fossil energy and nonrenewable material inputs with green energy and locally sourced resources (Chap. 2) and eliminating or recycling pollutants. The key to these processes at the social and economic level is to be able to create new economic opportunities and collective stewardship within local communities, growing the value of tourism and other tradable local services while maintaining patterns of biodiversity on land and sea.

These steps and their articulation are illustrated in Fig. 10.1. Detailed plans associated with each measurable, specified over time and space must be developed for each island system.

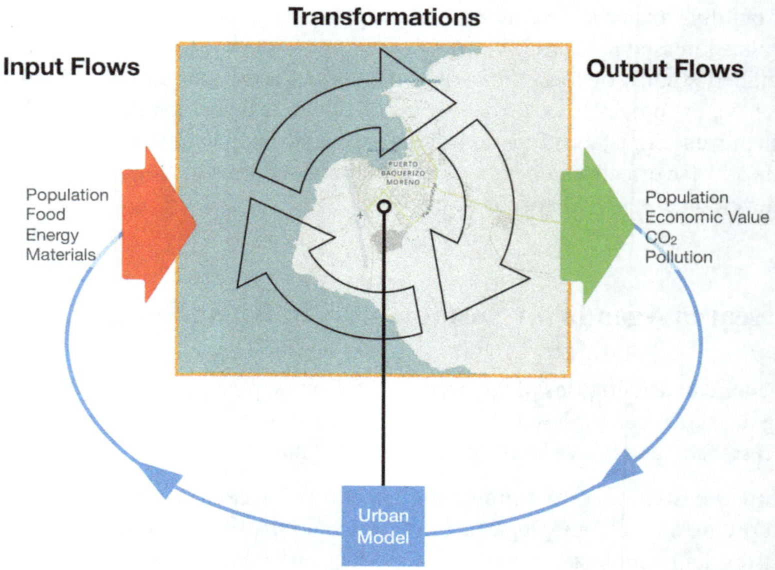

Fig. 10.1 Input and output flows of human activity in an Island system (Puerto Baquerizo Moreno, San Cristobal, Galapagos). The transformations of inputs (red arrow) into outputs (green arrow) are described in terms of a model of the urban settlement (blue box) specified in terms of dynamical networks of people, infrastructure, and places. The return blue line connecting outputs back to inputs via the model represents potential changes in the integrated operation of the system toward sustainability

Outlook

There is a critical need for a deeper, more operational understanding of processes of human development that are sustainable. The focus has recently shifted from global effects, such as CO_2 in the atmosphere, to sources and control points, such as the localized processes where energy is consumed and where unsustainable consumption and waste production take place.

Island systems present a perfect opportunity for creating true sustainable dynamics of human development. This is not only because these natural systems are unique and therefore irreplaceable – a fact long known to evolutionary biologists – but also because their human settlements are fairly small and far less complex than megacities elsewhere. Because of their physical separation and relative simplicity, island urban systems can now be mapped in terms of their structure and flows in an increasingly complete way; see Fig. 10.1.

Island urban systems are challenging in other ways that make a local solution play to higher standards relative to more contiguous geographic settlement systems. Their physical isolation strongly suggests that most input resources and energy should be sourced locally and that likewise all wastes should be processed in place so that they are not harmful to the environment. These constraints depend on how sustainable transportation in and out of islands becomes, a process that is also undergoing transformations (McConnell 2002; Ng and Song 2010; Upham 2003), and on relative advantages of production and post-consumption processes in islands versus in other locations.

Going forward, it will become critical to carefully lay out and map these sustainability scenarios, especially those in which the flows of production and consumption underlying the (tourism) economy of islands such as the Galapagos are to be closed either locally or more globally, if transportation itself becomes sustainable.

While it remains true that the battle for sustainability will need to be won in cities, it may well be that it is in coupled urban-natural systems in islands (Chap. 3) that the most critical breakthroughs will be made. Making economic growth and sustainability mutually supportive in island urban systems is therefore a critical – more tangible and more exacting – step toward realizing global sustainability goals everywhere.

References

Acemoglu D, Johnson S, Robinson JA (2005) Institutions as a fundamental cause of long-run growth. In: Handbook of economic growth, vol 1. Elsevier, Amsterdam NL, pp 385–472

Barro RJ, Sala-i-Martin XI (2003) Economic growth, 2nd edn. The MIT Press, Cambridge, MA

Bettencourt L, West G (2010) A unified theory of urban living. Nature 467(7318):912

Crutzen PJ (2006) The "anthropocene". In: Earth system science in the anthropocene. Springer, Berlin/Heidelberg, pp 13–18

Darwin C, Mayr E (2003) On the origin of species. Harvard University Press, Cambridge, MA, 18. print., a facs. of the first ed

Darwin C, Browne EJ, Neve M (1989) Voyage of the Beagle: Charles Darwin's journal of researches. Penguin Books, London/New York

Ellis EC (2011) Anthropogenic transformation of the terrestrial biosphere. Philos Trans R Soc Lond Math Phys Eng Sci 369:1010–1035

Ellis EC, Ramankutty N (2008) Putting people in the map: anthropogenic biomes of the world. Front Ecol Environ 6(8):439–447

Isard W (1972) Location and space-economy: a general theory relating to industrial location, market areas, land use, trade, and urban structure. MIT Press, Cambridge, MA

Jacobs J (1985) Cities and the wealth of nations: principles of economic life. Vintage Books, New York

Jones CI, Romer PM (2010) The new Kaldor facts: ideas, institutions, population, and human capital. Am Econ J Macroecon 2(1):224–245

Kennedy C, Pincetl S, Bunje P (2011) The study of urban metabolism and its applications to urban planning and design. Environ Pollut 159(8):1965–1973

MacArthur RH, Wilson EO (2001) The theory of island biogeography. Princeton University Press, Princeton

McConnell M (2002) Capacity building for a sustainable shipping industry: a key ingredient in improving coastal and ocean and management. Ocean Coast Manag 45(9–10):617–632

Ng AK, Song S (2010) The environmental impacts of pollutants generated by routine shipping operations on ports. Ocean Coast Manag 53(5–6):301–311

North DC (1987) Institutions, transaction costs and economic growth. Econ Inq 25(3):419–428

Romer PM (1990) Endogenous technological change. J Polit Econ 98(5, Part 2):S71–S102

Sahely HR, Dudding S, Kennedy CA (2003) Estimating the urban metabolism of Canadian cities: Greater Toronto area case study. Can J Civ Eng 30(2):468–483
Upham P (ed) (2003) Towards sustainable aviation. Earthscan, London UK
Whittaker RJ, Fernández-Palacios JM (2007) Island biogeography: ecology, evolution, and conservation, 2nd edn. Oxford University Press, Oxford/New York

Index

T. Kvan, J. Karakiewicz (eds.), *Urban Galapagos*, Social and Ecological Interactions in the Galapagos Islands, https://doi.org/10.1007/978-3-319-99534-2

CPSIA information can be obtained
at www.ICGtesting.com
Printed in the USA
LVHW082142221118
597953LV00003B/14/P

9 783319 995335